PRODUCTS THAT LAST

ACKNOWLEDGEMENTS This book is the result of the research project *Products that Last*. The authors gratefully acknowledge the support of the Innovation-Oriented Research Programme 'Integrated Product Creation and Realisation (IOP- IPCR)' of the Netherlands Ministry of Economic Affairs, facilitated by the Netherlands Enterprise Agency (RVO).

The authors would like to thank our business partners in the *Products That Last* project consortium for their active cooperation and contribution to our research: Ahrend Produktiebedrijf Sint-Oedenrode, Beroepsorganisatie Nederlandse Ontwerpers (BNO), Enviu, Interface Nederland, Océ Technologies - A Canon Company, Koninklijke Auping, PARK Noordeloos, Philips Electronics Nederland and Vodafone Libertel as well as those companies outside the consortium that helped shape our thinking by allowing us to interview their key people or working with us in student graduation projects. A special thank you goes to the Chair of our project committee Jos Oberdorf (npk design) as well as the RVO programme managers Michiel de Boer and Christien Dohmen and RVO programme consultant Joop Postema.

We would also like to express our thanks to all our TU Delft IDE students, who during their graduation projects or our lectures both challenged and supported us.

IN ALL THINGS THERE IS A LAW OF CYCLES - REBUS CUNCTIS INEST QUIDAM VELUT ORBIS -

PRODUCT DESIGN FOR CIRCULAR BUSINESS MODELS

Conny Bakker · Marcel den Hollander · Ed van Hinte · Yvo Zijlstra

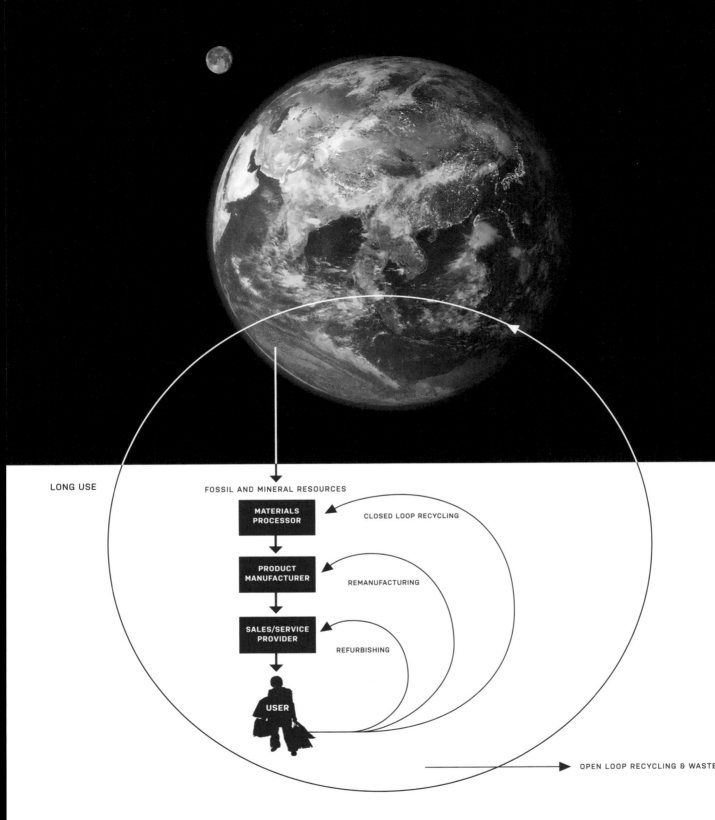

LONG USE

FOSSIL AND MINERAL RESOURCES

MATERIALS PROCESSOR

CLOSED LOOP RECYCLING

PRODUCT MANUFACTURER

REMANUFACTURING

SALES/SERVICE PROVIDER

REFURBISHING

USER

OPEN LOOP RECYCLING & WASTE

When viewed from a sufficiently large distance, it becomes evident that both product life extension and recycling are different articulations of long use, the first intervening at product level, the latter at material level.

FOREWORD

A throwaway culture has characterised and underpinned Industrial economies for more than half a century. Only recently, however, have governments and researchers begun to pay attention to the short lifespan of many consumer goods, prompted by concern at their wastefulness and a growing awareness that their embedded carbon has implications for climate change, demanding a slower throughput of materials in the economy.

Too many goods do not last as long as they could – or should. Some fail, others become unwanted. Certain types of product have declined in quality as companies have cut costs to remain competitive. All too often consumers' purchasing decisions have prioritised style over substance. Poor design and high labour costs have caused repair and upgrading to be fringe activities.

Most businesses operating in our growth-dependent economic system, in which success is judged by ever-increasing sales volumes, have remained silent in this debate. They fear that if products were to last longer, sales (and thus profit) would inevitably decline. Such logic needs to be questioned: environmental sustainability does not require a decrease in the value of consumption, only in its volume, or weight.

It has become clear that many companies will need to change how they operate if they are to survive while the throughput of products in the economy is reduced. The 'pile them high and sell them cheap' business model is no longer credible. Profit must in future be generated through improvements in the intrinsic quality of goods and enhanced after-sales services.

This will be no easy task; it demands change in people's mind-sets and a willingness to take risks. Products That Last makes a valuable contribution to understanding a vitally important debate and identifying the business opportunities ahead.

Tim Cooper
Professor of Sustainable Design and Consumption, Nottingham Trent University
October 2014

CONTENTS

PREFACE

The urgency to develop industry, trade and consumption towards a sustainable dynamic has become all too familiar. These days we discuss Circular Economy as the solution to our troubles and recycling is considered by many as the knight in shining armour who will make that happen. However, against the background of making products last, Circular Economy is an important tool, but the technical question if a material can be shredded and regenerated is only a rather trivial part of it.

This book is certainly not about recycling. That issue is indeed addressed in the book complementary to Products that Last, entitled 'Products that Flow' (also BIS Publishers). It concerns the manageability of flows of fast moving goods, such as disposables and packaging and relevant business and design potential.

The questions we get asked about lasting goods concern equally profound systemic shifts. We have to move from creating things that are produced, sold and mostly neglected to become waste, towards making products with value that we can cultivate during longer periods of time. That is what is implied by changing from a linear to a circular economic model. The difficulty here is not the environmental preferability of a circular over a linear economy, for there is no way on earth in which we can continue selling things and ignoring the consequences over the next 50 years or so. The difficulty does lie in finding ways to get there. How should we shape the circular economy? What kinds of products, services and schemes should we be developing?

This book is the result of a three-year research effort into business scenarios that prolonged product lifespans may present. It helps slow down the speed of the flow of materials and goods through society, addresses consumption practices, reduces waste and 'buys us time' for more careful development.

Systemic change is not easy, but certainly tempting. The challenge is to find different perspectives, which at first may be unusual, for business development and fresh design concepts to match. The key is to analyse what your current tacit assumptions are and use those to envision long term opportunities. Making products last requires exploration of what may happen long after they were put to use. Think 30 years. Designers are able to explore new directions and express ideas that are in line with new business models. Entrepreneurs can focus on redefining their responsibility and on developing new partnerships that can contribute to sustaining long term product value. At the end of the day circularity first and foremost depends on cooperation.

We therefore dedicate this book to all active change agents and open-minded people: join us. Let us begin by challenging the 'sell more, sell faster' habit and explore the opportunities offered by a much longer product life.

Conny Bakker, Marcel den Hollander, Ed van Hinte, Yvo Zijlstra

Image left: We have produced 8,000 tonnes of space debris, which basically means some 29.000 objects which are bigger than 10 centimeters and, maybe, around one million smaller ones. Most of them will live forever. Collisions with new objects will multiply their number and increase the danger of more accidents.

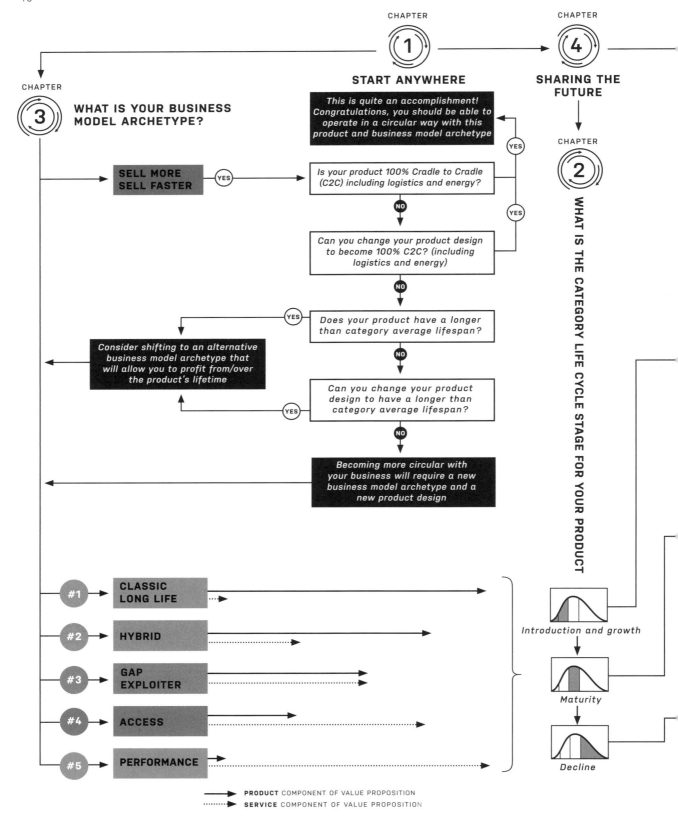

CHAPTER
(1)

START ANYWHERE

CHAPTER
(4)

SHARING THE FUTURE

CHAPTER
(3) **WHAT IS YOUR BUSINESS MODEL ARCHETYPE?**

CHAPTER
(2)

WHAT IS THE CATEGORY LIFE CYCLE STAGE FOR YOUR PRODUCT

This is quite an accomplishment! Congratulations, you should be able to operate in a circular way with this product and business model archetype

SELL MORE SELL FASTER — YES → Is your product 100% Cradle to Cradle (C2C) including logistics and energy?

NO

Can you change your product design to become 100% C2C? (including logistics and energy)

NO

YES → Does your product have a longer than category average lifespan?

Consider shifting to an alternative business model archetype that will allow you to profit from/over the product's lifetime

NO

YES → Can you change your product design to have a longer than category average lifespan?

NO

Becoming more circular with your business will require a new business model archetype and a new product design

#1 → **CLASSIC LONG LIFE**

#2 → **HYBRID**

#3 → **GAP EXPLOITER**

#4 → **ACCESS**

#5 → **PERFORMANCE**

Introduction and growth

Maturity

Decline

→ **PRODUCT** COMPONENT OF VALUE PROPOSITION
┈┈▸ **SERVICE** COMPONENT OF VALUE PROPOSITION

CHAPTER **5** **WHICH PRODUCT DESIGN STRATEGIES APPLY TO YOUR PRODUCT?**

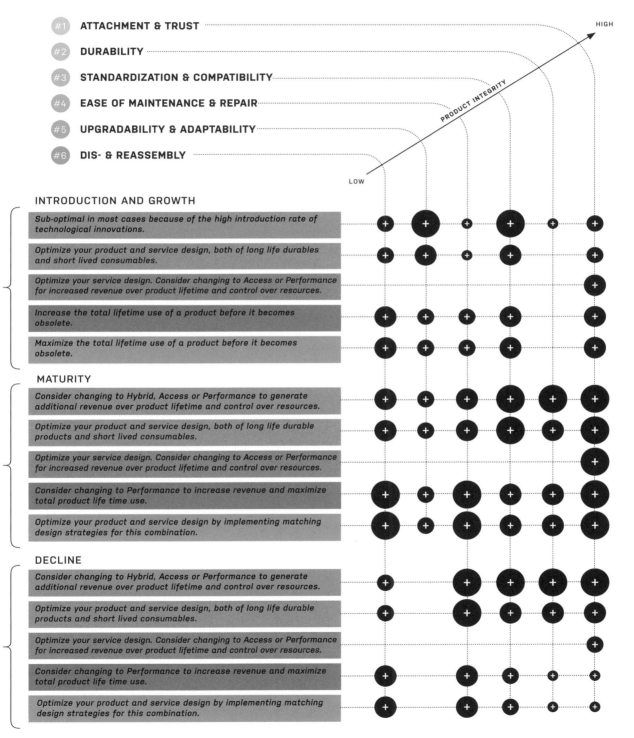

#1 **ATTACHMENT & TRUST**

#2 **DURABILITY**

#3 **STANDARDIZATION & COMPATIBILITY**

#4 **EASE OF MAINTENANCE & REPAIR**

#5 **UPGRADABILITY & ADAPTABILITY**

#6 **DIS- & REASSEMBLY**

PRODUCT INTEGRITY

HIGH

LOW

INTRODUCTION AND GROWTH

Sub-optimal in most cases because of the high introduction rate of technological innovations.

Optimize your product and service design, both of long life durables and short lived consumables.

Optimize your service design. Consider changing to Access or Performance for increased revenue over product lifetime and control over resources.

Increase the total lifetime use of a product before it becomes obsolete.

Maximize the total lifetime use of a product before it becomes obsolete.

MATURITY

Consider changing to Hybrid, Access or Performance to generate additional revenue over product lifetime and control over resources.

Optimize your product and service design, both of long life durable products and short lived consumables.

Optimize your service design. Consider changing to Access or Performance for increased revenue over product lifetime and control over resources.

Consider changing to Performance to increase revenue and maximize total product life time use.

Optimize your product and service design by implementing matching design strategies for this combination.

DECLINE

Consider changing to Hybrid, Access or Performance to generate additional revenue over product lifetime and control over resources.

Optimize your product and service design, both of long life durable products and short lived consumables.

Optimize your service design. Consider changing to Access or Performance for increased revenue over product lifetime and control over resources.

Consider changing to Performance to increase revenue and maximize total product life time use.

Optimize your product and service design by implementing matching design strategies for this combination.

Products that Last flow diagram: Questions for identifying the right business model and product design strategy.

START ANYWHERE

14

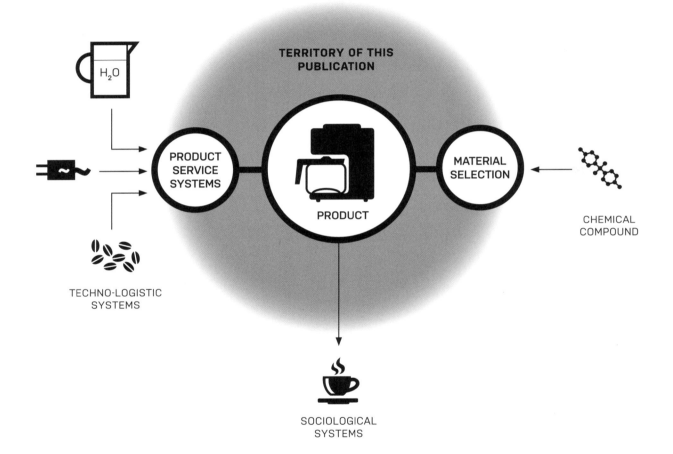

TERRITORY OF THIS
PUBLICATION

H_2O

PRODUCT
SERVICE
SYSTEMS

PRODUCT

MATERIAL
SELECTION

CHEMICAL
COMPOUND

TECHNO-LOGISTIC
SYSTEMS

SOCIOLOGICAL
SYSTEMS

The product domain is wedged between product service systems and available material choices. This book is about interventions within the circle of influence of designers and businesses.

Some people own two functioning coffeemakers and a bunch of broken down ones stashed away somewhere. Sooner or later they will discard the wrecks, possibly when clearing up storage space. Or a functioning machine decides to start leaking and the owners decide to replace it with a brand-new and improved coffeemaker.

LONG TERM FRAMING

This ongoing family ritual represents a minute part of a stream of mass production, consumption and destruction of zillions of goods. So, what's new? Now imagine a different setting, one in which people lead a comfortable life without the constant noise of making and breaking products in the background. Companies focus on long-term convenience by offering products with lasting value and making a profit through exploiting and sustaining those products. This requires a drastic change towards entirely different ways of doing business and, consequently, a shift in what designers should focus on. The easy part is that the basics for alternatives have been around for ages. It is just that never before were they regarded as ways to engage in trade with a view to the extension of product life and saving materials and energy. *Products that Last*, provides a range of business models and design strategies with examples and backgrounds. It intends to serve as an inspiration to make the change happen. It is essential, therefore, that the subject matter of Products that Last is kept simple and that the overview provided is a clear one.

It concerns a shift of emphasis in the way entrepreneurs and designers develop and exploit goods towards reduction of material and energy consumption over time. Nothing more, much less.

The effort to develop new ways of doing business is part of a sizeable and varied attempt to turn our currently fragile life style into one of which we can be fairly certain that we can sustain it well into the remote future.

Products that Last proposes useful methods and strategies. There are no generic rules. Each case requires its own specific recipe. It offers a different perspective on commonly used notions in the sustainability discourse. 'Life cycle' is probably the most important concept in the book. There are various categories of life cycles. The kind aimed for in Life Cycle Assessment, for example, represents environmental impact. In the book product life cycle mainly concerns value changes. The emphasis is on maintaining value.

Energy consumption is only partly relevant to product lifespan extension. It is not a significant element when it concerns the design of a specific electric dishwasher, because value is the main issue. But it is certainly meaningful when the evolution of dish washing enters the picture. As technology develops over time, energy consumption per product is likely to decrease. The optimal product lifespan then is defined by the point in time where the environmental impact that arises from using a product equals the embedded impact of a - more energy efficient - replacement product. So, in some cases, early replacement can be an eco-effective strategy.

A similar evolution occurs in devices that produce and store energy, such as windmills, solar panels and batteries. For those it is important to note that they have a lifecycle too, in the 25 years range. This implies that their value is subject to cultivation as well. Free energy is so attractive that this fact is sometimes overlooked.

Using minimal amounts of material through thinking in lightweight structures has an aim similar to product lifespan extension. The former implies a direct reduction, the latter a reduction through intensified use. Both are in line with the adage of visionary designer Richard Buckminster Fuller: 'Do more and more with less and less until eventually you can do everything with nothing'. The many dome structures he designed are a close approximation. The future of technological development has a lot in store when we combine value cultivation with a diminished use of materials.

Perceived product value over time only partly depends on functional properties and the scenario of material interventions needed to keep the product going. Perceived meaning, the immaterial, STORYTELLING element is equally important for maintaining its reputation. In practice this may concern branding, advertising and organising what could be called 'image enhancing activities'. Up till now these efforts have been carried out exclusively within the linear model of design for sales, with the exception of the odd 'lifelong guarantee' advertisement. There are opportunities here that may affect the development of business models.

The last domain to be mentioned is HISTORY of design and commerce. Very little evidence is available regarding the relationship between assumptions about design and product longevity. The problem with the design of lasting products is that it must start out from speculation, and that it will by definition take a long time to be proven right. Historic longitudinal research could be of great value here. There is no future without learning from the past.

EXPIRATION DATES The epitome of linear product development is of course planned obsolescence, which is the opposite of any attempt to make products last. The principle dates back to the early 1920s, when an annual change of model in the bicycle and car industries was proposed as a way to nudge customers to keep on buying the latest model. In 1932, Bernard London introduced the expression "planned obsolescence" - as a way to reanimate the economy in the US after The Great Depression. His particular interpretation might even have justified his proposals. Industrial designer Brooks Stevens, on the other hand, popularised the term "planned obsolescence" in the 1950s specifically as a means to make customers purchase the latest designs. Commercial critic Vance Packard distinguished between London's emotional obsolescence and the rational functional kinds, where products either break down after a predefined period of time or become too costly to be used economically.

Mythology has long surrounded this issue, and this was commented on by Dutch writer Karel van het Reve, who worked as a newspaper correspondent in the Soviet Union in 1967 and 1968. He wondered why Western industry benefited from the limited lifespan of light bulbs, whereas the light bulb lifespan was much shorter in Russia, where those involved in production couldn't care less about repeat sales. He also observed something very peculiar: people in the Soviet Union were prepared to pay for light bulbs that no longer worked. As it turned out, they collected them to swap them for working light bulbs at the office i.e. get a functioning light bulb when you need it, without having to wait for state bureaucracy.

The abandoned house of the Bulgarian communist party.

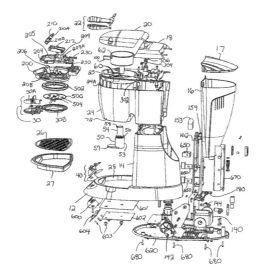

To improve your own life by improving someone else's is the standard individual engine of economic activity. Services are performed that range from producing and selling huge container vessels with enormous quantities of trade in their economic wake, to buying one packet of paper handkerchiefs a day and selling them one by one, thus making them affordable to the poor. You can also choose to attend to a sick person in an intensive care unit, teach children, lend money, sell reconditioned engines for Korean cars, or offer shower cabins for rent on a tropical beach. In other words, people offer value for which other people are prepared to pay.

VALUE THE OPPORTUNITY

The world of values is overwhelming and chaotic. One way to provide an overview is to regard it as a festive mix of value systems and choose one as an example. We have picked the traditional electric coffeemaker, because it represents all the models and strategies described in *Products that Last*. Apart from being a popular functional object, it consumes energy and needs other products - coffee and filters or pads/cups - to do its job.

Let us start from the beginning: the design. The first element in creating value is non-material. A small group of people are paid to envision a new coffee-making device with its very own characteristics in order to distinguish it from its competitors. Since electric coffeemakers have reached their maximum functional potential, the added value is projected mainly through form, texture and colour. The new design addresses a certain lifestyle stereotype, and it is not exceptional that, from a functional quality point of view, it is

inferior to its predecessors. The prototype and the plan to produce and market the item represent the first layer of added value.

After the design and optimisation stage comes production, the most obvious process for achieving product quality. Apart from the material devices used and the factory lay-out as value adding services in their own right, this requires raw materials: metals, plastics and bonding agents in a preproduction form - granulates, sheets, tubes, fluids and so on. Producing these materials from ore, oil and vegetation, including the transportation effort required adds more value. However this, paradoxically, combines with a loss of value, because getting raw materials into shape requires energy. And there will inevitably be waste. You win some, you lose some.

This phenomenon also applies to the production and assembly of parts in order to manufacture this provider of steaming hot drinks. The amount of production waste can be considerable. It usually includes

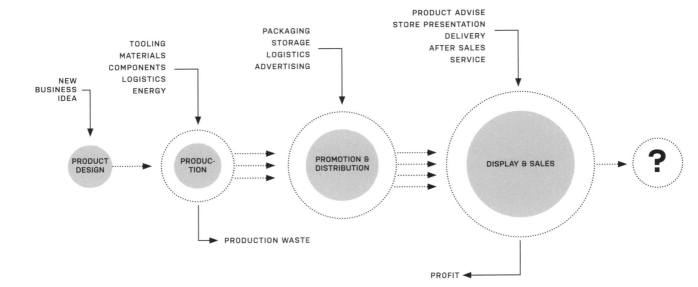

In a linear economy value is added in each step, but after sales the product tends to disappear beyond the 'newness horizon'.

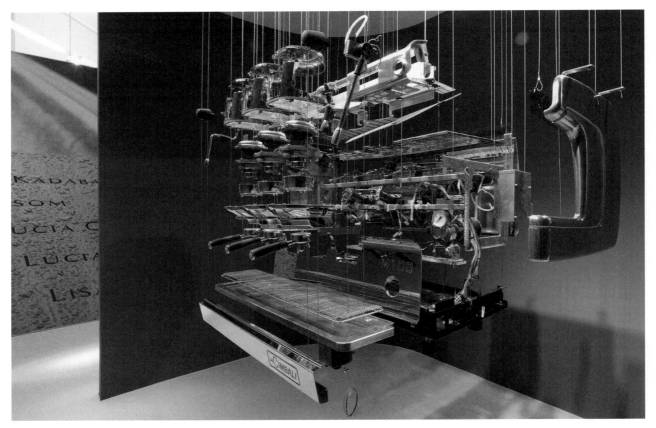

Last coffee espresso machine of Gruppo Cimbali - M100 - on display at MUMAC, the Museum of Coffee-Making Machines near Milan, Italy.

Low-tech ways to make coffee: the perculator, the ROK espresso maker and the French press.

superfluous materials that will become useless, or raw materials which will be fed to recycling machinery to regain value.

Finally the brand new coffee machine, together with seemingly relevant printed matter, is packed in a cardboard box which protects it and advertises it on the outside. It is transported with numerous of its fellow products on trucks and ships, briefly stored, put on display in a shop or at a market, and sold to a customer. All these actions provide extra layers of value.

The final value factor is what the customer is prepared to pay on top of the actual cost which, incidentally, includes what the producer and suppliers were prepared to pay. That final factor constitutes the income generated by the new coffeemaker for the people involved in designing, producing and marketing it.

Here we arrive at the traditional horizon of linear product development. Beyond this line of 'newness', product lifespan considerations are, in general, virtually absent. From the perspective of the producer and the designer, after it has been sold our poor coffeemaker ends up in the 'coffeemaker hereafter': and it hasn't even brewed a single cup of coffee.

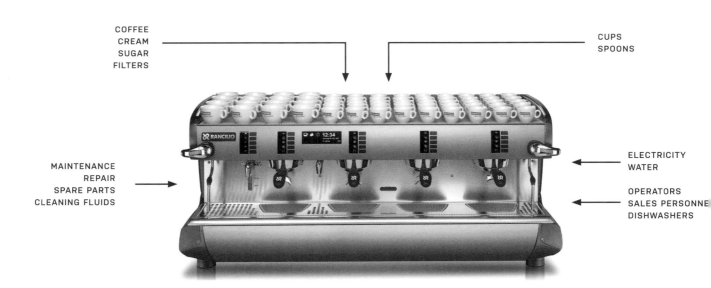

COFFEE
CREAM
SUGAR
FILTERS

CUPS
SPOONS

MAINTENANCE
REPAIR
SPARE PARTS
CLEANING FLUIDS

ELECTRICITY
WATER

OPERATORS
SALES PERSONNEL
DISHWASHERS

A new design of a lifestyle stereotype is not necessarily a functional improvement.

By no means, however, does this imply the end of value creation. Other parties become involved, all intending to make a profit from what is already there and being used in homes, on the streets, in offices and other workplaces. For a start, there are companies that grow coffee beans and multinationals that buy them, transport them, process them and sell coffee of different brands and qualities. Next, coffee needs to be dispersed and dissolved in hot water that must pass through a filter, for which there is also a supplier. The appliances need maintenance and repair. Some may need cleaning fluids and spare parts provided by specialised firms. Although it has never been explicitly investigated, it is likely that a considerable part of the world population makes a living by servicing what is already there.

Apart from the suppliers of extras, there are companies in the trade of catering and event organising that rent out coffeemakers. Of course there are also businesses that provide coffeemakers plus maintenance to offices and other workplaces. These hot drink machines may be more robust and larger, but they are suitable as an illustration of the world of coffeemaker value creation. And there are even more businesses involved with coffeemakers.

After a number of operational years, coffee devices wear out and become redundant. At this point they are either stored in an attic or a shed for later use (which will never happen), or taken to a shop for refurbishment to retrieve at least some of their value. They are resold as bargains to perform in extra time. But alas and inevitably, their end is nigh. With a few unique exceptions of coffeemakers that once brewed coffee for celebrities or have otherwise gained access to the 'heaven of antiques and expensive stuff', they end up in the flow of waste. They will be taken apart and some parts might make it to the repair shop once again. Nevertheless, the coffeemaker's destiny is to be partly scrapped and recycled, and partly burned to provide some consolation energy. Though scrapping and burning do serve to salvage some of the original value, considerably more is destroyed as part of these inherently destructive processes.

All these value-changing transactions happen beyond the newness horizon. There is a certain awareness of this next universe, but up till now it has pretty much been taken for granted. This is caused on the one hand by a combination of prosperity and a habitual human fascination for all that is new, and on the other hand by the fact that, in the current situation, entrepreneurs and designers tacitly assume that more money

can be made along the road to new propositions. There have been very few practical attempts at unravelling product afterlife and systematically scrutinising it for opportunities.

Opportunities clearly do exist, or they would not right now provide an income for so many people. They need reframing, however, to be recognised as segments of a circle of continuous value creation. Lifespan is not something to be passively observed after products have been put on the market, but involves a cluster of possibilities to be taken into account by entrepreneurs. Together with designers, companies can work on developing new principles of making money through the use of products: lifespan implies a progression of possible steps to add value.

Becoming familiar with this new perspective on product development requires cooperation between producers, designers and professionals in current post sales services, depending of course on the kind of trade. Coffeemakers belong to a different culture than washing machines, or bicycles, or smart phones, or novelties. Entrepreneurs and designers can explore markets accordingly.

With a different focus ideas for products and services are bound to change. There will be a shift in emphasis on various product characteristics and the concept of service development will gain importance. Products will be almost like people, expecting and getting life improvement.

Confederate snare drum from the Battle of Glorieta Pass (March 26-28, 1862) at the New Mexico History Museum, Santa Fe. Acoustic musical instruments are meant to be maintained and repaired and usually outlive their makers. Before the digital revolution this was also true for cameras. Analog cameras used to be lifelong companions. Digital equipment on the other hand, is difficult to adapt, refurbish or repair. It tends to be outdated within a few years.

Millions of choices and plenty of opportunities to get everything.

Some samples of what contemporary amateur photographers can afford. Do they need it? No. Do they want it? Yes.

Drumming without a kit. Aerodrums works with reflectors, light, a camera and a computer to make them work. It was created for the musician who wants to be expressive and versatile while solving the problems of portability, space and noise associated with regular drum kits.

MY NEW CAMERA My digital camera is six years old. It still works pretty well, apart from the little flap that should open and close when the telescopic lens comes out and goes in. I have two batteries, one of which cannot be trusted. The optical quality is great. Resolution in new cameras, however, is considerably higher.

However, I see a desirable camera of a different brand, with better zoom capacity, plus my income is improving and I'll be going on holiday soon. That pushes me over.

So I start looking at the newest cameras of my preferred brand and there is an even better one. I decide that I want. I can afford it, but I decide I'll opt for the cheaper earlier version. Then I go to the camera shop. There I learn that there is only one left, at a different branch. And it is white. That does it: I purchase the latest camera.

UTILITY FASHION During WW 2 materials became hard to come by. The UK government started a Utility Fashion scheme. 'To save material, unnecessary pleats, double breasted jackets, and long socks for men were not allowed. Buttons, buckles, zippers, clasps and elastic were limited because metal and rubber were needed for the war industry. Skirts were knee-length, coats were shortened, and boys under thirteen were not allowed to wear long trousers. (..) Colours, however, were not restricted. Fabric prints for dresses for example were often small, busy patterns of colourful flowers that were easy to put together without having to waste fabric to make the motif connect. The Board of Trade estimated that millions of meters of cloth were saved by all of these regulations.' (Marjanne van Helvert, The Responsible Object, Valiz Publishers, Amsterdam 2016).

HER NEW INSTRUMENT My eight-year-old granddaughter has been taking drum lessons for over a year now, so the choice for a birthday present was obvious. A musical instrument stimulates acquiring fine motor skills as well as knowledge and experiencing the joy of music.

Ambition is the only limit apart from the budget. This last requirement led to choice for a second-hand. Was it going to be a colourful rock kit or a digital one? Browsing revealed that almost all rock kits are black. The world of rock-'n-roll was less colourful than imagined. A digital one is ideal for smaller apartments with thin walls, because it is compact and one uses headphones. Still, as soon as she masters this kit she'll want an acoustic one, because it sounds so much richer and is more adventurous: pure magic. And it doesn't need current!

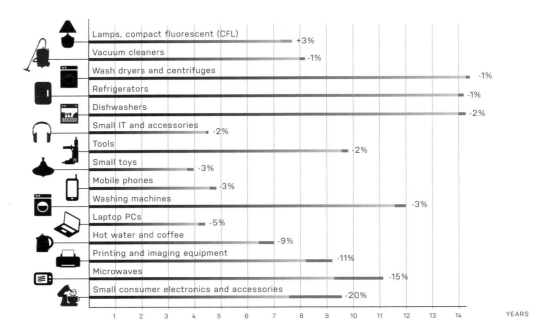

Median lifespan of a selection of household products, and change over time (2000 – 2005), based on Dutch data. Source: Bakker, C.A., Feng Wang, Jaco Huisman, Marcel den Hollander (2014) Products that go round: exploring product life extension through design. Journal of Cleaner Production, 69, 10-16.

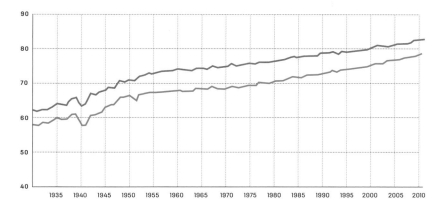

Life expectancy at birth in the UK, 1930-2011, female (green) and male (red) Source: Office for National Statistics, UK.

A strong rationale underlies the extension of the lifespan of a product. There are three domains where we can find arguments in favour of long-life products. These stem from human characteristics and the observation that the reduction of the ongoing product lifespan collides with the increasing scarcity of raw materials.

LOADS OF REASONS

The first domain regards the ambiguous affinity of humans with their 'stuff'. It depends on age and individual mentality, and even on situation and mood. Generally people crave for the new. They follow the latest fashions and acquire things for practical or symbolic reasons, or just for the fun of it. At the same time, however, they cherish the familiarity of objects and rituals, are interested in the past and may indulge in nostalgia. Irish comedian Dylan Moran said it simply, 'there is too much of everything, and not enough.'

The second domain concerns business people and designers who seek to learn how to make a positive contribution by taking up new challenges. This is not so much a preference for developing and producing objects, but rather an inclination to observe, evolve and apply their skills, and to make a point.

The third and most important set of reasons emerges from awareness of the unintended side effects of what we have accomplished, as a rapidly expanding species that not so much adapts itself to living conditions, but rather continuously adapts its environment to what it figures it may need. Up until recently, little attention has been paid to the depletion of materials and the fact that so much is wasted as a result of the way production and consumption have developed.

Let us start, rather philosophically, with the domain of human idiosyncrasies. Life is not easy. After birth we need maintenance and care to survive for a considerable amount of time before we become old age pensioners and, finally, die. Without this care and maintenance we wouldn't be able to survive for the number of decades we almost seem to take for granted nowadays. Maximising our lifespan in a state of reasonably good health requires a great deal of dedication and loads of stuff. Excluding basics like food, family and clothing, here, in random order, is a list of a number of human resources and implements that a certain female X requires to help maximise her lifespan: Spectacles, nurses, scissors, buggies, band aids, painkillers, teachers, tweezers, blankets, oxygen, toothbrushes, ointments, internet access, sunglasses, crutches, counsellors, hearing aids, glasses of water, mouthguards, doctors, books, antibiotics, beds, therapists, splints, mobile phones, sneakers, dentures, pads, soap, dentists, walking sticks, serum, handkerchiefs, plaster, stairlifts and towels. This list illustrates that we consider it appropriate and go out of our way to single-mindedly protect, maintain and repair our bodies and minds.

Our attitude towards products is far more ambiguous. Care is not obvious. On the one hand, we may feel it is

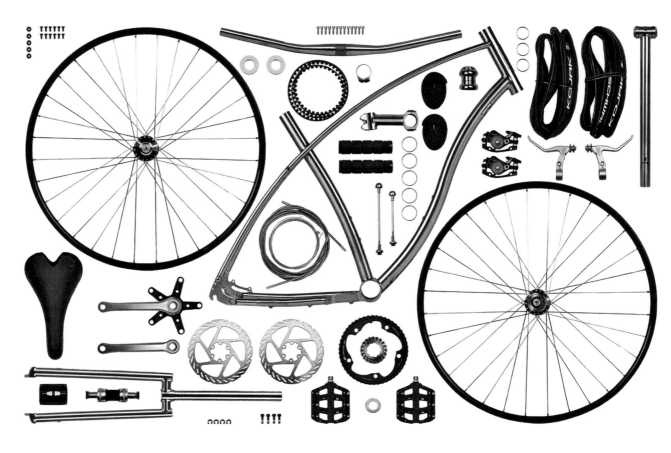

A bike is a beautiful example of a modular system that is easy to repair. The parts are standardized and come in a range of different colours and qualities. In theory it is possible to buy each part separately and assemble a totally custom bike. While the basic design of a bicycle hasn't changed much over the past century, parts like disc brakes and belt drives are still being improved to increase durability and facilitate maintenance (Images by Budnitz bicycles, budnitzbicycles.com).

a shame if something we have used for some time has to 'die' because it fails to fulfil our needs any longer, and we may consider letting it live' for a bit longer. On the other hand, the purchase of a 'new and improved' successor promises a satisfactory experience. The decision to buy is quite complicated and depends on a great number of things. The point is that in general most products are expected to last, just like people. It is not exceptional that the first reaction to failure is to consider repair or some other measure, to restore the product to an acceptable level. Deep down one wants to hold on to it, as long as it doesn't draw the wrong kind of attention. Here we have an intriguing seesaw with, on the one end, the traditional one way trip from production to garbage, with a bit of excitement somewhere along the way, and on the other, the development of ways to create enduring products, and cause a shift in what people spend money on. Lasting products are preferable.

The decisive edge to shifting the seesaw towards sustainability rests within the second domain, and can only be obtained through the development of new kinds of trade, to which end designers and producers will need to understand the principles. This constitutes the second domain: taking up the challenge.

This book gives you the basics. In addition, it is important to become aware of what makes trade tick; it may very well be the same force that drives consumption: convenience. Making a profit and satisfying your investors quickly and with little effort is essential. The other requirement is that your business is based on a credible challenge which is also exciting and fun to tackle.

Products with a long lifespan can offer a wide range of business opportunities, depending of course on the type of product and its design. Cars are a good example - to a certain extent. They are traditional providers of after-sales business: they need continuous repair and maintenance and, they can be rented, shared and leased. This implies more intensive use, which represents a different view on lifespan extension, and may invoke different quality requirements and therefore a different design outcome.

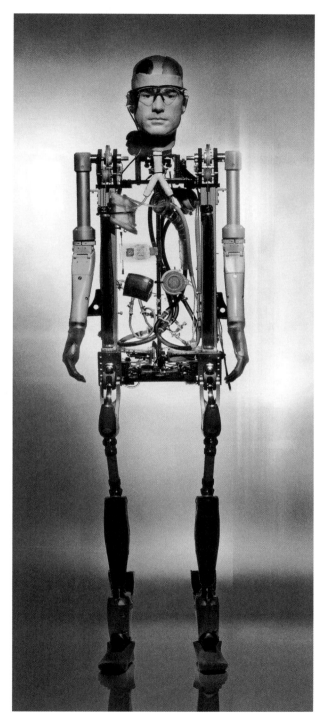

Rex, the world's first bionic man, modeled after University of Zurich psychologist Bertoit Meyer has synthetic blood, artificial organs and robotic limbs. Retinal and cochlear implants allow him to see and hear. All of his components could theoretically be welded to a human body to replace missing or worn out parts.

Hiroshi Fuji, 'Where have all these toys come from?' The exhibition took 'kaeru' as its theme (meaning to change, to return, to exchange) and brought together over 50,000 toys gathered over 13 years of Japanese toy exchange systems.

The comparison with less self-evident examples such as coffeemakers, furniture, digital information devices, pianos, ladders, microwaves, skateboards and even knickknacks – to name but a few, is likely to provide ideas for existing after-sales business, but could be 'internalised' into systematic design and responsible entrepreneurship. Consequently, these products could be improved on to extend the lifespan of a product. In other words, observing examples might, with a little imagination, produce arguments for lifespan extension in the guise of interesting questions. Making products last is a challenge.

The third domain is the best known. Mankind produces and consumes too much 'stuff'. It has become an addiction, which is a consequence of being trapped in the vicious circle of cheap production to satisfy mass markets and being forced to sell vast numbers to be able to produce cheaply. Because of this, the number of products sold needs to grow continuously.

Part of this can be considered as compensation for population growth (which also has its limits) and emerging trade, but sooner or later successful markets will become saturated. When that happens, a limited lifespan becomes an economic necessity, because replacement purchases are all that remain. From then on, new and cheap will reign.

There are three ways in which this economic pressure has its effect. The first, obviously, is that what little attention there was for product lifespan will be reversed. This is the direct result of the guiding principles of reducing and externalising costs in traditional linear business and product development: design something, manufacture it at the lowest possible cost, sell it at the highest possible margin and forget it as soon as feasibly possible. Secondly, products age in relation to the context. If they happen to contain elements that improve very fast as a result of clearly aimed technological development, their overall ageing speed will increase 'against the will' of some of the more steadfast components. It is no revelation to state that the development speed of electronics is so high that it is virtually impossible to purchase anything really new in that field. 'New and improved' is already looking over your shoulder when you are in the middle of a purchasing decision. Finally, in the third place, there is deliberate product ageing, in style or in functionality: planned obsolescence, as mentioned in the previous chapter.

Although it is not usually subject to scrutiny, there are indications that a product's lifespan is shorter than before. A survey of a range of standard electrical household products in the Netherlands between 2000 and 2005 showed a lifespan decrease in all but one, which happened to be the compact fluorescent lamp - a relatively new item which has a reputation for longevity to live up to. The lifespan of small consumer electronics decreased by 20 per cent, which can partly be explained by the abovementioned high-speed development of digital technology.

The observed lifespan decrease also coincided with considerable economic growth seen at that specific period. It is not improbable that since the credit crunch in 2008, acquisition of replacements has become less affordable to most people in the West. It is quite likely that product lifespan correlates to economy, but this subject has not been explicitly explored.

Continuously increasing production, particularly combined with the growing demand for commodities in emerging large economies and their ability to produce quite cheaply for the rest of the world, inevitably affects the economic availability of material resources. Prices, on average, have gone up and - even more detrimental - have become more volatile. Therefore, retaining value throughout a product's life through careful management is the sensible thing to do.

Riversimple's Rasa; a hydrogen car that is not for sale but can only be leased on a monthly full service contract

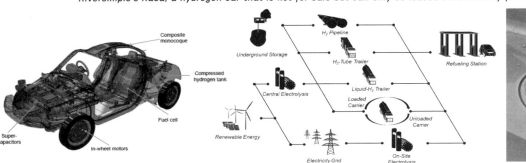

Unlike hydrogen cars, electric cars rely on heavy batteries. Lithium is one of the main resources needed for car battery production. The Salar del Hombre Muerto district is a part of the so-called lithium triangle which is believed to be home to half of the world's lithium reserves. Bolivia wants to be a relevant actor in the global lithium market and started huge mining operations on it's giant salt flats.

Rendering cars environmentally harmless is particularly difficult, with all these identity carriers moving back and forth, each a considerable inert mass. An exemplary attempt to achieve that is the Rasa (from the Latin 'tabula rasa', or empty slate), developed by Riversimple in Wales.

MOBILITY VALUE

It runs on hydrogen. Yet the most important point the company wants to make, is that in general transport efficiency is the backbone of its mobility concept, which specifically for the Rasa happens to entail the use of a fuel cell. Exploiting hydrogen as an energy accumulator for a simple automobile is the option they demonstrate. The company mainly focuses on increasing vehicle lifespan and keeping its weight to a minimum.

The car is the ultimate commodity. It is technically, socially and psychologically as complicated as products can get. Also, it is a familiar phenomenon to every world citizen, both functionally and symbolically, in use and through advertising. Cars are often depicted in a straightforward lie, as the trustworthy mobile safe haven in a vast empty environment. Several brands photographically even made it to Mars, where traffic jams are unlikely to happen. Moreover, cars are subject to all business models discussed in this book.

Currently, the car is on the brink of a major breakthrough to comply with changes in environmental insights and requirements. This entails a shift in car perception from identity symbolism, a steel apparel image, towards true transportation functionality.

The current push towards vehicle electrification is getting stronger every day. Converting electrical power into movement is more efficient than burning fuel and using the heat to drive an engine. Electrical power is clean too, as far as individual vehicles themselves are concerned: no CO_2 pipe at the back. What is most likely to change in conversion and storage before e-power actually reaches storage in cars in batteries or in hydrogen, or perhaps some other medium remains to be seen.

The most popular option so far, is to stuff a car with batteries that can simply be charged from the power grid. The disadvantage is that one battery easily outweighs two passengers and, of course, needs to be transported. In theory one can extend the range of

a car by adding more batteries up to the point, where the vehicle becomes too heavy to move an inch. Ergo, for a given car concept there is likely to be an optimum range defined by the choice between reducing body weight and increasing battery weight.

Producing batteries in sufficiently vast amounts and sustaining their value on top of that, without causing ecological damage, is challenging to say the least. Production of sufficient renewable power may be unattainable if growth of energy consumption is assumed to be unavoidable.

Riversimple's Rasa two seater weighs 580 kilogram. A full tank of hydrogen, from a local station, adds only 1.5 kilogram, nevertheless providing it with a maximum range of about 485 kilometres with a smart system of distributing and recuperating power. A fuel cell generates it and braking sends surplus power back to capacitors. The cell is only used when cruising. Acceleration and driving uphill are powered by the capacitors. Fuel consumption amounts to the equivalent of a staggering 88 kilometres per litre. This description does not yet account for the production of compressed hydrogen. The other form in which hydrogen functions as energy carrier is liquefied. Hydrogen atoms can be attached to liquid molecules from which they can be 'picked' again with the help of a catalyst. The Australian Israeli company Electriq boasts a recyclable (undisclosed) water based hydrogen fuel that can be replenished with fresh hydrogen. The assumption behind both hydrogen systems and also behind battery systems is, that electricity eventually will become renewably sourced. If this will actually happen is unclear. A process of transition can be accelerated by drastically increasing transport efficiency. In that sense Riversimple is on a promising trajectory.

This route includes the Rasa business model. These cars are not for sale. One pays a certain amount per month and everything is taken care of, including insurance and even fuel. Here we have a typical Access model, in which the producer is the keeper of responsibility for everything, except driving.

RE: The fact that verbs concerning product lifespan extension usually start with the prefix 're-', vaguely suggests a certain reluctance, as expressed in, Oh no, not again!' Nevertheless, there are standardized (British Standard BS8887-2-2009) definitions for these words:

REPAIR: returning a faulty or broken product or component back to its usable state.

REFURBISHMENT (RECONDITIONING): returning a used product to a satisfactory working condition by rebuilding or repairing major components that are close to failure, even when there are no reported or apparent faults in those components.

REMANUFACTURING: returning a used product to at least its original performance with a warranty that is equivalent or better than that of the original product. A popular expression for a colourful kind of reconditioning is 'pimping'. Maybe 'repairing' could simply become 'mending', A straightforward 're-less' alternative for 'refurbishment' is 'maintenance'.

They all serve towards reuse of products. Recycling is any recovery operation by which waste material is reprocessed into products, materials or substances, whether for the original or other purposes.

Some designs become classics, like the wonderful Volkswagen van. Some VW affectionados like to keep their vans operational and in mint condition while purposely showing and cherishing the patina of aging. The aesthetics of aging can thus become a fashion statement, with stonewashed jeans as an extreme example. The aged jeans look is achieved with sandblasters and bleach, resulting in an artificially aged product with a considerably shortened lifespan.

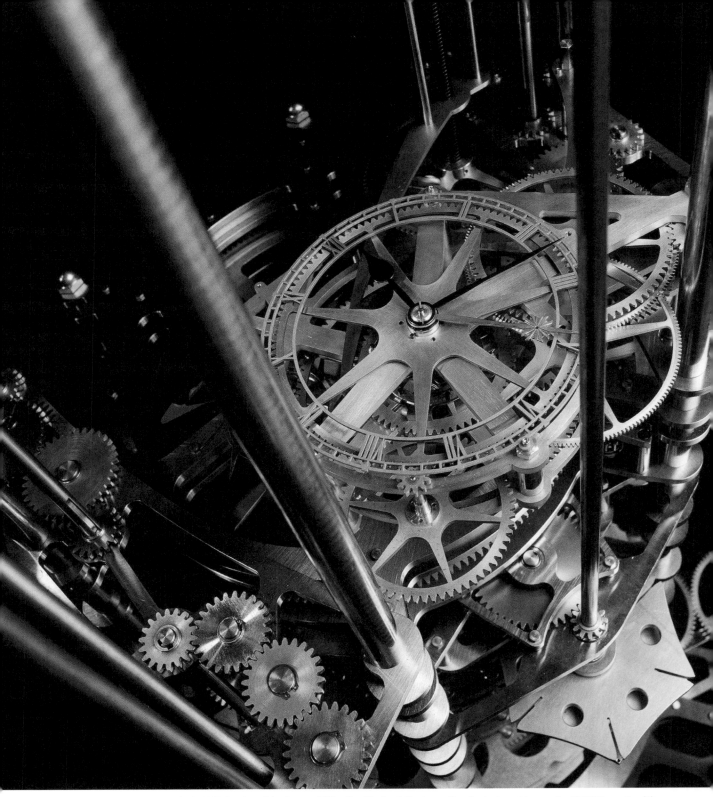

THE CLOCK OF THE LONG NOW *Designed by Danny Hillis, the Clock of the Long Now is designed to run for 10,000 years with minimal maintenance and interruption. The Clock uses the energy captured by changes in the temperature between day and night and will also be wound up by using the energy generated by its visitors. The primary materials used in the Clock are marine grade 316 stainless steel, titanium, high-tech ceramics and stone. The entire mechanism will be installed in a remote limestone mountain in west Texas. The picture shows a prototype of the Clock.*

In circular models of production and consumption, all matter and energy is dealt with responsibly. Models always represent ideal situations and processes; reality is more fuzzy and dirty. Nevertheless, if we keep in mind which underlying assumptions we use, models can help us consider better ways for production to fulfil consumer needs.

ROUND AND ROUND IT GOES

A model starts with the perception of reality: by drawing a diagrammatic model, you take a few steps back, like a painter does, to view the complete picture. The awareness that, traditionally, production and consumption are organised as a linear process with a beginning and an end, requires this overview to make us realise that alternative setups are possible. By viewing a map representing all the stages, functions and identities of, for example, the coffeemaker, other options may emerge.

The bottom line is the notion of 'ecology'. Ellen Swallow Richards was the first to coin the word in 1892. Since then its meaning has evolved, from 'home economics' to 'the study of interaction between organisms and their environment'. Through ecological analysis, mankind arrived at the realisation that life forms and their environment are intertwined. It is possible to identify ecosystems - situations, areas, regions, or the whole world if you like - in which elements and substances go round in circles, or loops. Ecology works when loops are closed. Ecological systems can sustain themselves only if they manage to close nutrient loops. Such closed loop systems are regarded as sustainable (in the sense of being self-sustaining over long periods of time). They do, however, need an external source of energy which in nature is sunlight. Currently in human societies it is oil – which of course makes any attempt at developing a closed loop system unsustainable from the start.

Another important point worth noting concerning the nature of these circles is that they merely indicate repeating qualities. Water fulfils a sequence of functions in circular order: it drenches land, feeds plants and land animals, is home to marine life, absorbs part of the power of the sun in clouds, falls as rain, and so on and so forth. Each particular water molecule, however, will always follow a different circular path. In the same way proteins, calcium and a host of other substances and elements go through cycles. They are all part of life, over and over again.

Materials that originate from forestry and agriculture also consist of these same elements and substances. The same system that produces them breaks them down into their basic components. They are harvested and characterised by the fact that they fit in a circular organisation of transformation in which they play a range of roles of which, for some substances, 'being edible' is probably the most important property for human beings. This kind of transformation is spontaneous. It takes effort and energy to grow and exploit trees, plants, fungi and animals, but they return to their value adding potential literally by their nature.

When this system approach is applied to technical materials and structures, the outcome is far less relevant to ecology. From the outset, the focus in production is on adding considerable value, often in complicated ways, so much so that it takes a great deal of energy and effort to restore complete products to their original raw material potential after they have done what they were supposed to do. Closing the loop here is a matter of economics and is therefore not necessarily ecological. Because the economic value added is, apart from functionality and ownership, mainly meant to make products briefly stand out among the crowd by advertising their newness, their destiny is landfill, or being recycled at a relatively early stage of their useful lives. They were not designed for long-term value. There is no loop, just a straight line from the cradle to the grave.

To illustrate potential change, the butterfly diagram developed by the Ellen McArthur Foundation can be a great help. It is designed to demonstrate the principle of circular economy - as opposed to linear economy - and it pictures the transformation circles of technical materials as a mirror image of the circles through which biological materials move around. Because of the focus on material loops, the butterfly model barely takes energy use into account; it does show energy recovery from burning materials at the bottom of the diagram, but energy input is absent. To add both

MAINTAINANCE

VALUE UPGRADES

VALUE UPGRADES

RECYCLING

The top of the cone represents the highest value of a product. Materials and products should be designed and treated in such a way that they remain as close to the top as possible for a long period of time, and that the 'falling speed' of value loss from the top to lower circles is minimal. Photo's: CRH380 harmony bullet at a Wuhan high-speed train maintenance base, China (above) and Uyuni Train Cemetery, Bolivia (below).

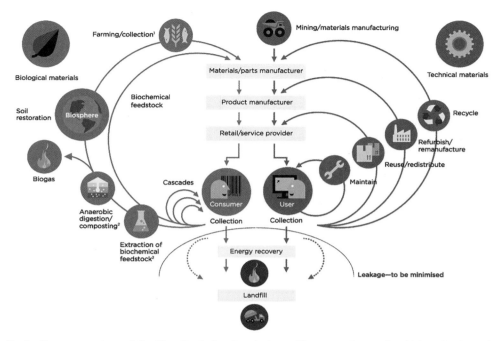

Butterfly diagram, courtesy of the Ellen MacArthur Foundation. It illustrates the way in which technological and biological nutrient-based products and materials circle through the economic system.

ecological and economic value, energy is indispensible. Therefore maintaining value - and not destroying and reconstructing it - saves energy.

The diagram, particularly the technical side on the right, can be perceived as a hierarchy of value treatment circles in different processes. If the wing on the right (technical materials) were three-dimensional, it could represent a cone in relief, or a mountain, with maintenance circling around the top of the hierarchy, recycling around the bottom, and value upgrades in between.

The idea of a hierarchy clarifies the crucial point that materials and products should be designed and treated in such a way that they remain as close to the top as possible for a long period of time, and that the "falling speed" of value loss from the top to lower circles is minimal, or even negative, in which case value increases. In theory this can be achieved by favouring maintenance and reuse over refurbishing and remanufacturing until recycling, which is a radical value reducer, to restore material potential to its first stage, is unavoidable. Time provides a set of requirements here: loops will have to be stretched, or

the amount of use per unit time will have to increase, to be able to speak in terms of products that last.

When considering the above, the butterfly diagram suddenly starts to reveal the differences between various kinds of circles. Some represent closing economic loops, others the closure of ecological circles, and still others represent both. This hidden distinction is paramount. In an economic system, closing the loop is about business maintaining control and preserving economic value of assets over their lifetime. In an ecological system, however, closing the loop is ultimately about safely reducing man-made materials to nature's own chemical compounds, which in many cases means the opposite for businesses: relinquishing control and annihilating substantial economic value.

Consider for example recycling synthetic plastics, often used as a more general example of 'closing the loop'. Closer examination reveals that we are actually dealing with a special case of "closing the loop" in an economic system. It is an attempt to preserve value or reduce cost by making the material last longer. A manufacturer can achieve this through either repurposing - application in a different context - or repair - adding virgin plastics to make the material meet the original specifications. Thus, the ecological loop, contrary to popular belief, is not closed at all.

Protocell material combines non-living chemicals that self-assemble to behave like living cells.

Designer and researcher Shamees Aden developed a prototype running shoe (the Amoeba trainer) out of Protocell material. The trainers are 3D printed and customized to fit like a second skin. Synthetic biological Protocell material enables the running shoes to self-repair. Mixing different protocells creates different properties with different behaviours, depending on heat, light, and pressure.

The advantage of recycling is that the reproduced material can be used for different products than the ones it originated from. It is a new start: what once was a PET bottle can now become a sweater or a car part. On the other hand, reuse of the complete water heating system of an old coffeemaker in a new one is more favourable - no value destruction - but also more complex, since it requires a high level of "organised predestination". It is precisely this way of thinking that deserves more attention.

To close the ecological loop for bioplastics, there would be no need to return them to a manufacturer after end-of-life disposal, as they would effortlessly and of their own accord disintegrate into nature's own chemical compounds. In the economic sense, however, the loop would remain fully open in this scenario, since there is no end-of-life collection of materials and therefore no control over material life extension or value whatsoever.

The circular economy concept strives to be both an ecological and an economic closed loop system. Pointing out the difference between them shows that for product lifespan extension, the focus is on building and maintaining economic value. The only advantage of biodegradable materials is that value destruction is free.

It is important to note once again that at the end of the day, value is always determined by the market. A brand-new Ferrari will easily lose 100,000 Euros within its first two kilometre spin, simply because almost immediately after delivery it will be 'secondhand' to a potential buyer. A restored painting by Mondrian on the other hand, which will always at best be 'secondhand', may become more valuable because of the publicity it gets when purchased by some billionaire. The impli-

cation of these extreme examples is that, ultimately, economic value always requires customers. A remanufactured coffeemaker needs someone who wants it, regardless of what it went through in previous cycles.

There are more considerations concerning the butterfly. The diagram loosely distinguishes between two families of materials, the biological ones and the technical ones. Industrial practice, however, involves products that consist of many different materials that can belong to either family. The other complicating fact is that products or parts made out of biological materials can also be susceptible to repair, maintenance, reuse and even recycling outside the 'bio-loop'. For example wood chips from milled production waste can be combined with a synthetic resin and thereby become 'technical'. So depending on how they are used, biological materials can behave like technical

The cobbler maintains shoes and upgrades their value. Well-made quality shoes pass through many circles and may outlive their wearers.

materials, the main difference being that they do not need any costly help to return to their organic origin. Technical materials will never be able to behave biologically, no matter how hard they try.

This observation points the way to a further refinement of the butterfly diagram: value treatment and treatment circles do not need to be the same for all product parts. For example, it is quite common that new trucks are sold with overhauled engines, which in turn may partly consist of new parts. In the future, this may be taken one step further in the sense that some part of product X may have been designed to continue life as a part for product Y. Mercedes signs are virtually designed to become bling accessories. This is of course an obvious example. The point is that value transformation loops mingle, within as well as between products. For business and design, this represents a great challenge.

BMW V8 engine

ETERNAL CYCLES

The 'Hoodoos' rock formation at Bryce Canyon began life some 144 million years ago. For 60 million years a great shallow sea built this area, depositing sediments as it repeatedly invaded, retreated, and then re-invaded the region. It left sediments thousands of feet thick. In the Tertiary Period, between 66 and 40 million years ago, the highlands eroded into shallow lakes. They deposited Iron-rich, limy sediments that grew to become the red rocks from which the Hoodoos were (and still are) carved. Frost wedging and gravity are the forces that create these fantastic shapes. Different rock types erode at different rates, causing the undulating landscape expression. In millions of years these great monoliths will crumble down to sand, to rise again like a Phoenix made of stone.

Wind turbines are supposed to produce energy efficiently for 12 to 25 years. Their function may also become obsolete for unforeseen reasons, leaving behind huge industrial installations that may become subject to some kind of gentrification.

Building and maintaining an energy infrastructure will always imply a lot of material, dedication, care and energy.

Polysilicon, crucial for the production of solar panels and turning sun radiation into electricity

Keeping a circular economy going requires energy. The current tacit assumption is that in due course all energy will be derived from renewable sources. It will therefore be available indefinitely, is supposed to be clean and it won't contribute to global warming.

CIRCULAR ENERGY

However, energy use is not a matter of availability, but rather of conversion, from one kind of energy to another, and storage. That is the job of energy transformers, smart logistics and storage media. The eagerness for what is known as 'the energy transition' is so strong, that a typical human habit hasn't been broken: ignoring the effects of equipment production and what happens after the deadline of delivery. Apparently for renewable energy we expect solace from future inventions just like we do with developments and discoveries around the use of fossil energy. In both cases we're 'borrowing' from future uncertainties.

Transformers of renewable energy, such as wind turbines and solar panels, are products with a limited lifecycle and value scenario. Wind turbines mechanically turn wind force into electrical power. To build them is not a clean process.

Particularly production of permanent magnets depends on rare earth element Neodymium, mining of which in China is an extremely dirty process. Currently there are German turbines with electro magnets installed and alternative permanent magnets are being researched. It is a shame that cleaner alternatives always have to wait their turn, until after environmental damage has been done.

Wind turbines are supposed to produce energy efficiently for 25 years. This is a crude estimation, since in practice they often only last for only 12 to 15 years, whereas with care and maintenance, function monitoring and repair they can be made to last longer. This is a typical matter of *Products that Last*: design and business need to reconsider options by thinking ahead much further than the currently foreseen life span. Wind turbines as a category may need to be kept going for decades, but their function may also become obsolete for unforeseen reasons, leaving behind huge industrial installations that may become subject to some kind of gentrification. Look at what happened to Dutch windmills, that used to contribute to water management and agriculture and are now considered watch food for tourists.

Solar panels suffer from similar caveats. Producers, most of them in China, do not have proper control of the poisonous waste caused by manufacturing polysilicon, crucial for turning sun radiation into electricity. Saving costs to be competitive leads to the cleaning up afterwards syndrome, which is the consequence of still naïve value allocation of current economics, in which pollution and landscape destruction are simply ignored.

The lifecycle of solar panels amounts to about 30 years, but may turn out longer, since power capacity drops by considerably less than expected.

The story of energy accumulation is more complicated since it involves devices of various kinds. The oldest is mechanical: save water behind a dam and consequently use the its weight to drive a mechanism. The second oldest is use the energy stored in certain fuels by burning. Batteries come next, as a way of storing electrical energy with chemical means. The fuel cell principle, which dates back to 1801 consists of producing electricity through a controlled reaction between hydrogen and oxygen. Hydrogen is a burning fuel, but also battery fuel. A fuel cell is a battery that is not charged by sending current through it, but runs on fuel instead.

Of all energy storage principles, the one with batteries is the heaviest per unit of energy. To avoid an overdose of detail, we limit ourselves to lithium-ion batteries. Their application is surging, particularly in e-cars. Production, prolonged use and recycling are very promising, but as yet unbalanced. Making these accumulators requires nickel, graphite, lithium and cobalt. Particularly the last two are hard to come by. Mining them is an expensive process and often involves child labour. This would imply that recycling batteries is favourable. In this life stage of lithium-ion batteries, however, the amount of batteries for recycling is limited and may remain modest for some time to come, because if their lifespan equals that of a car, there is likely to be as second life for them as passive energy back-up. This implies that including regenerated materials in new batteries on a scale that approaches the circular concept will not happen soon. Moreover, regulation of labour circumstances is called for, maybe in the sense that they get included in the 'efficiency' equation.

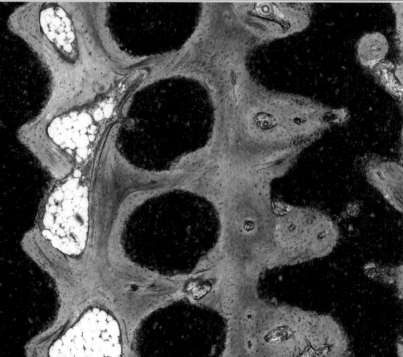

BIO-BUILDING Architecture team Softkill Design has developed a 3D printing technique for large scale construction which mimics the growth process of bone. A computer algorithm prints material into fibrous pieces that reinforce stress-prone areas. This creates a 'web' of material. Waterproof coating is applied inside.

BIO-LIGHT American biotechnology company Bioglow has applied synthetic biology processes to develop ornamental glowing plants. Starlight Avatar is the first plant that is able to light up autonomously, without the need for external treatments or stimuli such as chemicals or ultraviolet lighting. 'In the long term we see use of glowing plants in contemporary lighting design, namely in landscaping and architecture as well as in transportation, marking driveways and highways with natural light that does not require electricity. We also have a capacity to make plants glow in response to environmental cues, making them effective environmental and agricultural sensors.' Dr. Alex Krichevsky, Ph.D., MBA. Founder and Chief Scientist of Bioglow.

BIO-BRICKS With energy from renewable sources toy company Lego produces bricks from sugarcane-based plastic. They are 'technically identical' to pieces made of conventional plastic. They're not biodegradable, but can be recycled. It is a first attempt in the company's commitment of making Lego bricks sustainable.

BIO-BODY IMPLANTS Biomaterials and implants are frequently used to treat injuries and disorders of the musculoskeletal system to reconstitute function. Examples are osteosynthetic devices for the stabilisation of fractures, resorbable bone grafts and scaffolding for tissue engineering. Every biomaterial and implant should fulfil a specific mechanical function, be biocompatible, and even sometimes be able to influence biological processes.

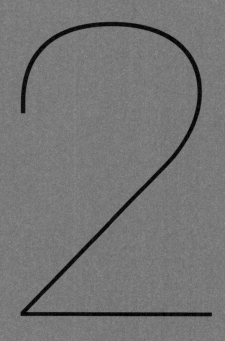

2

WHAT IS THE PRODUCT CATEGORY LIFECYCLE STAGE FOR YOUR PRODUCT?

PRODUCT CATEGORY LIFECYCLE CURVE

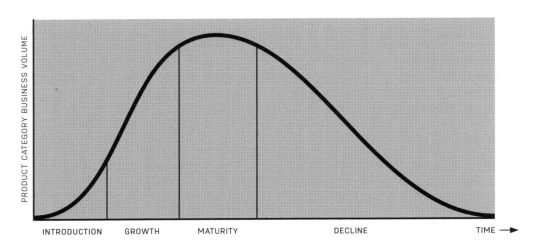

PRODUCT CATEGORY LIFECYCLE CURVE WITH EXAMPLES

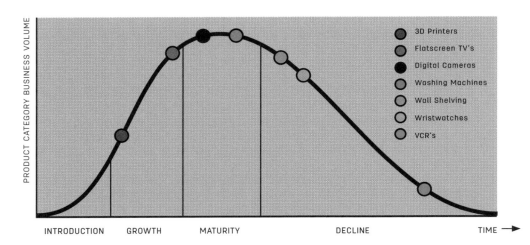

The most important beacon in developing business models and design strategies to make products last is the life cycle curve. If you administer and read it well, it shows you the stage at which your product is in the story of its life. It can even provide rules of thumb for adjustments, as pictured in the diagram at the beginning of the book.

UPS AND DOWNS

Paradoxically, the product life cycle curve is part of the legacy of the Sell More Sell Faster business model, because in itself it does not directly tell you anything about the life cycle of a product. It is not as straightforward as a reliability curve, which provides information about the probability of failure: high at the beginning, constant and acceptable in the middle and rising again towards the grim end. All that the life cycle curve shows is turnover in time for a certain category of products, which could be anything from a particular kind of peanut butter to a flatscreen TV. Up till now, companies have been using the 'bell curve' to check how products behave and to introduce a successor at the right moment.

This has no direct link with product lifespan as such… A limited edition expensive luxury Swiss watch could, for example, be sold over a period of just three months, resulting in a very short (and very high) curve. Nevertheless, the watches sold during this brief period could last as long as a century. The opposite is also possible. If the curve were to represent a tiramisu dessert, it could cover a period of 15 years, whereas in the fridge the dessert would keep for a couple of days at the most.

The curve represents the relationship between time and money received. At the first stages of the curve, introduction and growth, turnover tends to be low, although it does happen that certain products, such as tablets or game consoles, are sold out even before they have been manufactured (media make money by selling rumours). If all is well, turnover gradually increases until the second stage, that of maturity, is reached. At this stage, the product approaches the peak of its popularity. After having passed this peak, sales start to diminish, slowly in the beginning, faster for a while and finally more slowly again, resulting in a normal distribution pattern.

The term life cycle in itself does not concern actual product value over time, but rather the notion of business vitality. However, when we take the product life cycle concept one step further and consider the curve at product category level, it suddenly takes on a whole new meaning. Seen at this level, the stages become markedly different arenas for companies to pick their competitive battles.

The growth and introduction stages of a new product category are often quite dynamic. In the race for market share and brand equity, competitors tend to introduce a rapid succession of technological and functional improvements. Circular business models for this stage either enable users to keep up with the quick pace of change through clever and cost-efficient upgrades, or make the new technology accessible to a wider audience. Design strategies have to be chosen accordingly. It does not make sense to design a product using materials that can last 30 years, if it is to be expected that the product will be functionally obsolete after its first year on the market as a result of a competitor's technological innovation. Design strategies at this stage are most likely to focus on limiting the detrimental effects of frequent product releases on the average product lifespan. Design should allow for easy upgrading, disassembly and reassembly.

The camera lenses designed by Nikon are a good example. Nikon facilitated the transition from the old analogue 35 mm film to the new digital SLR technology, allowing customers to continue using lenses dating back as far as 1959.

PRODUCT CATEGORY LIFECYCLE CURVE WITH RELATIVE PRODUCT LIFECYCLE SHIFTS

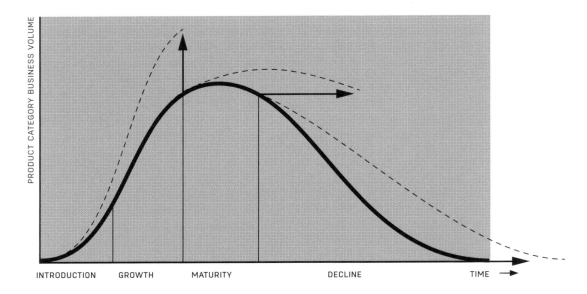

PRODUCT CATEGORY LIFECYCLE CURVE AND DESIGN STRATEGIES

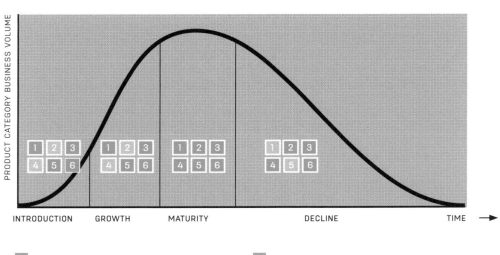

1 Design for Product Attachtment & Trust	**4** Design for Ease of Maintenance and Repair
2 Design for Product Durability	**5** Design for Upgradability and Adaptability
3 Design for Standardization and Compatibility	**6** Design for Dis- and Reassembly

Repair sits well with all four life cycle stages. During introduction and growth, innovative products are low in numbers but high in price. Early adopters will gladly pay for repair caused by inevitable teething troubles. In the maturity stage, prices drop, but the number of 'fallible' products on the market rises dramatically. This again is good for repair business. Finally, when decline takes over, product numbers dwindle and for those who want to hold on to their favourite things, parts and expertise soon become hard to find. Good news if your repair shop happens to stock those.

The maturity stage is less turbulent. In most cases, technologies will have reached an innovation plateau and changes are often incremental and cosmetic. Companies aiming to change their business model to become more circular will need to assess what business model archetype best fits their product portfolio, abilities, brand values, ambitions and their existing value chain. At this stage, striving for intrinsic product durability makes good sense and the full spectrum of design strategies for product lifespan extension can be successfully applied. Longer product life also enables the product to host value-adding services.

There are many examples of products and product categories at this life cycle stage. Miele washing machines that require Miele branded detergent cartridges, the Rolls Royce Performance business model, in which products are designed for durability, standardisation, and ease of maintenance and repair, and the Philips Lighting Performance model, which makes LED lighting commercially and financially accessible to a wider audience. To compensate for the drop in replacement sales (LED lighting lasts 20-25 years), Philips Lighting has introduced a full-service concept as a revenue stream. Refurbished iPhones, which happily coexist with newer models and keep giving access to Apple's iTunes and App stores, increase Apple's platform revenue and broaden their user base.

The downturn stage, finally, is characterized by dwindling product category revenue. At this stage, companies may decide to move out of the particular product category altogether or they may opt to keep operating in it for as long as possible. In the latter case, the decision can be made more profitable through design strategies such as cost reduction and providing high-end products, excellent service, durability and easy maintenance and repair.

Examples of products and product categories in the 'late maturity' or 'downturn' stage are Weller soldering equipment, with the sale of spare parts for almost all previous models, Vitsoe wall shelving and Le Creuset cookware.

The examples show that throughout the life cycle of a product category there are ample opportunities for businesses to become more circular. The maturity stage, which contains the majority of products we use in our daily lives, provides particularly fertile ground, in this respect.

An opened KR580IK80A Microchip - one of the most widespread Soviet processors. Contrary to popular belief, it appeared to be not an Intel 8080A (or 8080) clone, but a code-compatible redesign (while several parts are quite similar, routing is different as well as pad placement). Thinnest lines are 6μm.

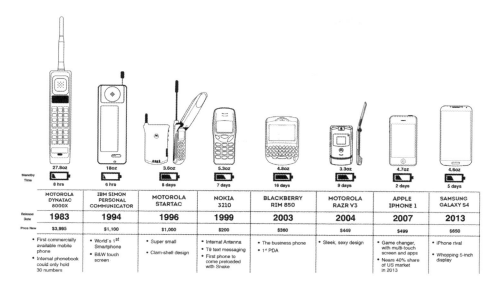

	MOTOROLA DYNATAC 8000X	IBM SIMON PERSONAL COMMUNICATOR	MOTOROLA STARTAC	NOKIA 3210	BLACKBERRY RIM 850	MOTOROLA RAZR V3	APPLE IPHONE 1	SAMSUNG GALAXY S4
Weight	27.8oz	18oz	3.6oz	5.3oz	4.8oz	3.3oz	4.7oz	4.6oz
Standby Time	8 hrs	6 hrs	8 days	7 days	16 days	9 days	2 days	5 days
Release Date	**1983**	**1994**	**1996**	**1999**	**2003**	**2004**	**2007**	**2013**
Price New	$3,995	$1,100	$1,000	$200	$360	$449	$499	$650
	• First commercially available mobile phone • Internal phonebook could only hold 30 numbers	• World's 1st Smartphone • B&W touch screen	• Super small • Clam-shell design	• Internal Antenna • T9 text messaging • First phone to come preloaded with Snake	• The business phone • 1st PDA	• Sleek, sexy design	• Game changer, with multi-touch screen and apps • Nears 40% share of US market in 2013	• iPhone rival • Whopping 5-inch display

Sustainability could do perfectly well without software. A conservative estimation of the amount of CO2 produced by data processing machinery for 30 'soft' Google queries is about the same as we emit into the atmosphere by boiling one 'hard' kettle of water. Mining crypto-money is already notorious for the disproportionate amount of energy it takes to produce monetary resonance on screens.

DATA THAT LAST

On the other hand, hardware has desperately needed software since Jacquard presented his punch card-controlled loom in 1801. Data storage and electronic computing devices are dead without programs. An unconscious smartphone is the epitome of powerlessness. Hard circuits without bytes moving around in them, are pointless.

This intimate relationship has mutual repercussions for soft- and hardware's respective life spans as well. The fast evolution of processors and memory, in size and data processing speed, increases programming code potential. The effect: growing size and complexity of software. As a consequence, it becomes too extensive for available memory. In user perception it also slows down devices that can be just a couple of years old.

Keeping software in line with expectations, maintaining it, getting rid of bugs, takes an awful amount of time, which consequently cannot be used for developing new software.

There is a rule of thumb that says that for any number of software developers involved in writing a given program, over half of them will be involved in maintenance within two years after introduction. Soon after that, only a fraction of the employed software engineers will be available for development of new programs. They can't keep up with improving technological potential anymore. The capacity to support existing software determines its longevity. Many digital products have an official (but not necessarily public) End of Life Date.

The cost of code maintenance is therefore, unfortunately, hampering 'backward compatibility', the capacity for up-to-date hardware to run elderly software. Simultaneously, new applications stimulate demand for higher processing speeds and more reliable, smaller and smarter hardware and connectors, the total disappearance of which seems to be the computing world's highest calling. In short: evolution of software and hardware are strongly interdependent, but not synchronous. They are like two dogs, each attached to one end of a leash.

Oak Ridge National Lab's Summit supercomputer became the world's most powerful in 2018, reclaiming that title from China for the first time in five years. The giant computer, , which occupies an area equivalent to two tennis courts, booted up a machine-learning experiment using more than 27,000 powerful graphics processors in the project. To train deep-learning algorithms it operates at a rate known in supercomputing circles as an exaflop.

Replacing machineparts by downloading 3D print backup files from the internet.

Recognized potential of new software applications creates new demand for hardware improvements. Some new developments directly relate to life span management, of all kinds of products, not just the obvious data processing kind. They, for instance, set the identity and monitor the functioning of hardware objects, thus contributing to the currently fashionable 'internet of things' notion.

Today, more than two decades after the first shuttle roared into space, booster testing still uses Intel 8086 chips, which are increasingly scarce. NASA plans to create a $20 million automated checking system, with all new hardware and software. In the meantime, it is hoarding 8086's so that a failed one does not ground the nation's fleet of aging spaceships.

Tagging products and parts, for instance with RFID (Radio Frequency Identification) circuits, facilitates tracking and tracing their whereabouts. This can serve different purposes, for example finding new or used parts to repair products, or assessment of the right recycling plant to deliver items, or to buy used materials that meet particular requirements.

Traceability can be enhanced and made more secure by distributing part data in a non-hierarchical database, a so- called block chain. Its principle is not very

Radio-frequency identification (RFID) uses electromagnetic fields to automatically identify and track tags attached to objects. The tags contain electronically stored information. Unlike a Replacing machineparts by downloading 3D print backup files from the internet barcode, the tag need not be within the line of sight.

complicated. This kind of database is not stored on one identifiable disk, or in one particular cloud for that matter. Instead information is spread across many disks in any number of computers owned by different participants. It is impossible to change data without involving others.

A centralized network (left), a decentralised network (middle) and a disttributed network. The last type is used for a blockchain database. Each participant maintains, calculates and updates new entries into the database. All nodes work together to ensure they are all coming to the same conclusions, providing in-built security for the network.

This kind of security is indispensable for protecting data. Within the book Products that Last, the perfect example of limited information accessibility is of course encrypting a 3D print model, so it can only be produced in a limited number per payment. Such a precise executable description is both the soul of a product and, paradoxically, an immaterial contribution to its individual sustainability: applying the description may take energy.

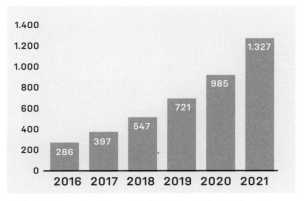

Global data center storage in Exabites. Globally, the data stored in data centers will grow 4.6-fold by 2021 to reach 1.3 ZB by 2021, up from 286 EB in 2016.

GENEROUS DISCLAIMER WITH A CAREFUL VENGEANCE

Product lifespan extension is not always the right way to make a positive contribution to sustainability. In general, new products are improvements, mostly in terms of energy efficiency, and at a certain point in a product's lifespan it may be favourable to replace it with a more efficient successor.

Contrary to popular belief, there is no fixed rule of thumb to establish when a product should be replaced. It depends on the product, of course, but more importantly on overall efficiency improvement speed. The product's performance is usually not constant and tends to diminish: improvement has its limits. It can be estimated through extrapolation, for example from the previous twenty years. Using that as a backbone, it is possible to build a model by mapping replacement scenarios (after one year, two years, three years, etc.) and calculating the effect on overall energy consumption. There is bound to be a moment in time at which there is a balance between replacing and not replacing. That is the optimum ecological lifespan (from an economical viewpoint the optimum may be different). After replacement, the same calculation can be made for the new product, and so forth.

In two case studies, assuming unchanged use of material, optimum lifespan estimations were made for a fridge and a laptop. A fridge currently has a twenty-year optimum lifespan. A laptop should last at least 10 years, but for this type of product the assumption of efficiency improvement with no material change is somewhat doubtful. Since the real average lifespan for fridges and laptops amounts to 14 years and 4 years respectively, they are both replaced far too early in their lives, causing damage to the environment. Here we have another argument for lifespan extension, or rather lifespan management.

Source: Bakker, C.A., Feng Wang, Jaco Huisman, Marcel den Hollander (2014) Products that go round: exploring product life extension through design. Journal of Cleaner Production, 69, 10-16.

Since 1965, Moore's law has ruled the computer hardware industry, doubling the processing power of computing equipment every 18 months. According to dr. Jonathan Koomey, computers have doubled in energy efficiency at about the same rate. The images trace this development, from the Electrical Numerical Integrator and Computer (ENIAC) that was completed in 1945 and is considered the world's first electronic computer to the AVIDAC, the first digital computer at Argonne National Laboratory that began operating in 1953; the IBM 360 mainframe (widely used in the 1960s); The VT100 video terminal, introduced in August 1978 by Digital Equipment Corporation (DEC) to modern energy efficient datacenters. Some applications, for instance in cryptography, may be able to become very much faster and energy efficient through quantum computing. Data are transferred via quantum teleportation based on the function of recently identified 'majorana-particles'.

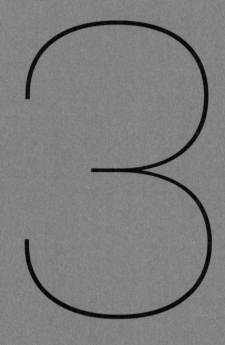

WHAT IS YOUR BUSINESS MODEL ARCHETYPE?

What is most essential, beneficial and profitable within the context of the tools we use, the skills we develop, the games we play, the chances we explore and the goals we aim for? It don't mean a thing if it ain't got that swing.
(Photo Harlod Eugene Edgerton, courtesey of MIT museum)

Making products last requires a revolution in ways of doing business. Making a profit from the fortunes of a product in the outside world does happen, but it needs an entirely different focus. The difference between current after-sales trade and the new concept of *products that last* is that in the latter case there is no deadline for a product to be definitely finished.

CIRCULAR BUSINESS

Money is made not so much through the sales transaction, causing the producer benefiting from products to lose interest, but through new business models designed to keep the wheel turning and where producers benefit from every turn taken.

Throughout its life a product can be modified, updated, repaired, sold, rented, borrowed, used for a range of purposes, taken apart, divided up, discarded, stored, rediscovered, filled and so on. Every change represents an opportunity to contribute to the closure of its economic circle, to create and maintain value over time, and to keep the revenue coming in.

The rationale of the processes driving a company is termed a 'business model', which includes everything that should be taken into account and dealt with for the company to create a value proposition that customers will accept and pay for. Theoretically, any business model can serve to make products last, provided it contributes to sustaining value and reducing material flows. In theory this even holds true for the conventional 'sell more / sell faster' business model. It is not inconceivable that high-speed sales of some ephemeral product, leaving no trace of value destruction, have the potential to reduce or even eliminate environmentally damaging turnover. Since the 'sell more / sell faster' model is not specifically meant for doing business involving products in use, let alone be applied as such, and because it is so familiar, it will not

be given further explicit treatment within the framework of circular business models.

There are several principles to try and apply when considering the development of a circular business model. The first principle, of course, is to aim for just that: thinking in circles. Next, it is vital to think in systems around products rather than just the product itself.

Consider combinations of products and services in alternative business models. There are always more: logistics, transactions, narratives, identity, you name it. These may well require cooperation with different partners, all responsible for different life cycle segments.

Leave products intact as completely and for as long as possible in all your scenarios, although it can be fun to play with different lifecycles and lifespans. Think about ways in which your customer or user can have new relationships and experiences with products, and ways in which they can use them.

Five archetypal business models benefit from a longer than average product life. The order in which they are presented is determined by the amount of service the product requires and by product type.

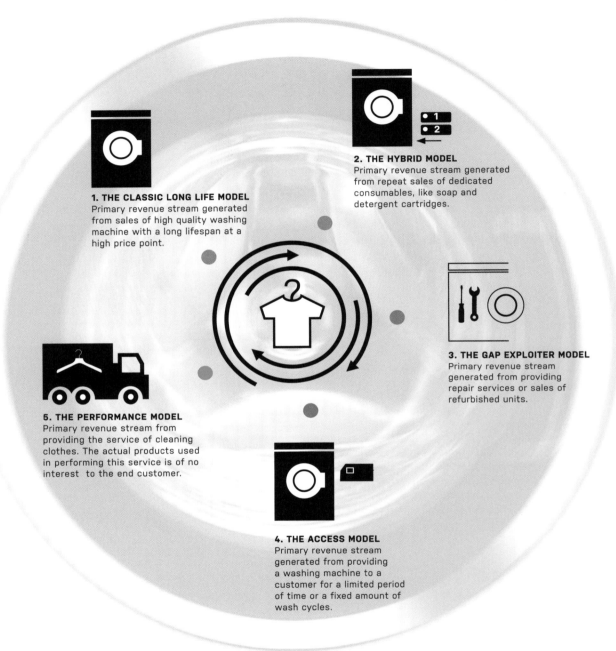

1. THE CLASSIC LONG LIFE MODEL
Primary revenue stream generated from sales of high quality washing machine with a long lifespan at a high price point.

2. THE HYBRID MODEL
Primary revenue stream generated from repeat sales of dedicated consumables, like soap and detergent cartridges.

3. THE GAP EXPLOITER MODEL
Primary revenue stream generated from providing repair services or sales of refurbished units.

4. THE ACCESS MODEL
Primary revenue stream generated from providing a washing machine to a customer for a limited period of time or a fixed amount of wash cycles.

5. THE PERFORMANCE MODEL
Primary revenue stream from providing the service of cleaning clothes. The actual products used in performing this service is of no interest to the end customer.

Five business model archetypes explained, using the washing machine as example.

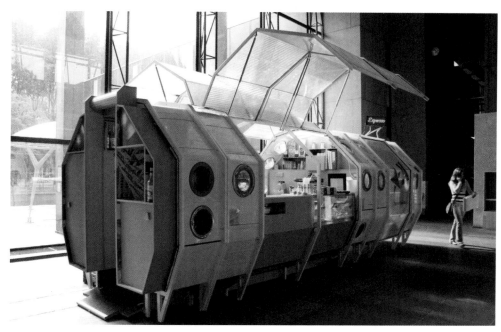

The Miele Space Station is an architectural installation made entirely of washing machines. The segments can be placed in various configurations, for various purposes and in different sizes. The space station has already performed as a bar, an art vending machine, a pavement cafe and a mobile architect's office. Source: Superuse Studios.

1. THE CLASSIC LONG LIFE MODEL proposes a high-quality product with a long lifespan, with sales as the classic source of income. After-sales support contributes to quality perception. In this model, brands more often than not have a reputation of good value for money, "it is not cheap, but it never fails".

2. THE HYBRID MODEL profits from the repeat sales of relatively cheap products with a short lifespan that only function together with a dedicated high-quality durable product. As examples, think of toner cartridges and coffee pads.

3. THE GAP EXPLOITER MODEL is interesting because it does not propose anything new, but feeds on value gaps in the existing system. It concerns the person in between, who repairs smartphones, sells second-hand equipment, turns CDs into candle holders, or whatever. eBay is a huge gap exploiter. Shoeshine-boys are small gap exploiters.

4. THE ACCESS MODEL is about making money through providing access to a product, while its ownership remains with the access provider. The customer gets to use a car (BMW), a washing machine (Miele), a room or a tool. There is a time limit and products are used in turns. The customer is concerned with the brand and with the kind of product.

5. THE PERFORMANCE MODEL leaves responsibility with the provider, with earnings based on the performance provided. It is up to the provider which machinery or products are deployed to carry out the required tasks. Users are exclusively interested in the quality of the service, not in the product providing it. Examples are printing, transportation or a data cloud. At work here is the so-called inseparability principle: services are consumed at the same moment they are generated.

What all these models have in common is the potential to maintain control of the flow of materials and products and generate profit from them over time.

BUSINESS MODEL CANVAS A 'Business model' is a typical example of an entity that is often discussed and used without a clear concept of what it actually is. Alexander Osterwalder and Yves Pigneur have developed a 'Business Model Canvas' that can be used to develop business models. In the words of the authors it, 'describes the rationale of how an

KP
KEY PARTNERS

What is the network of suppliers and partners that is needed make the business model work?

KA
KEY ACTIVITIES

What are the essential activities a business must do to make its business model work?

KR
KEY RESOURCES

What are the essential assets required to make the business model work?

VP
VALUE PROPOSITION

What value do we deliver to the customer?

Which one of our customer's problems are we helping to solve?

Which customer needs are we satisfying?

What bundles of products and services are we offering to each Customer Segment?

CR
CUSTOMER RELATIONSHIPS

What type of relationships does the business establish with specific Customer Segments?

CH
CHANNELS

How does the business communicate with and reach its Customer Segments to deliver a Value Proposition?

CS
CUSTOMER SEGMENTS

What are the definitions of the different groups of people or organizations the business aims to reach and serve?

CS
COST STRUCTURE

What costs are incurred to operate the business model?

RS
REVENUE STREAMS

What revenues are received from each Customer Segment?

The generic Business Model Canvas by Osterwalder and Pigneur (2010) subdivides organizational activities into nine building blocks.

organisation creates, delivers and captures value' in a fairly simple rectangular diagram consisting of nine 'Building Blocks', a term not entirely in line with the canvas metaphor.

The left side of the rectangle describes everything that has to be dealt with inside the complete organisation: create value. The right side describes all customer related activities: value capture. The value proposition is in the middle. As a shopping list template of the things you need to take into account, the canvas is indispensable.

Here is a very brief description of the canvas' building blocks. On the outer left we find the 'Key Partnerships' block. It sums up alliances and relationships with other organisations that contribute to your activities, which may concern anything from maintaining your strategy to the suppliers where you buy your nuts and bolts. In a circular economy, partners will include organisations with which value can be sustained in the full circle.

Next are 'Key Activities' and 'Key Resources'. The 'Key Activities' block contains all kinds of things your organisation needs to do, such as problem solving, production, communication, maintenance, monitoring, repair, etcetera, typically including value sustenance issues. The 'Key Resources' block describes what you need to have at your disposal to perform with. This can be physical (machines, infrastructure), intellectual (data, knowledge, patents, software, but also brand characteristics and - crucially - field information about your products), human (obviously) and financial (also obviously).

On the left is the money consuming side, so the fourth element, underpinning the previous three, is a description of the cost structure. The emphasis may be either on cost reduction or on value creation. Thinking circular may, for example, indicate that reducing material costs implies minimising material consumption, and that maintaining maximum value in respect of this modest amount of material is paramount.

Different kinds of approaches need to be chosen. Within a single business model it is possible that customers vary regarding distribution channel, relationship, profitability, value judgement, and even more. In addition, circular business thinking is likely to consider customer segmentation scenarios, as customer segmentation must be sustained over a long period of time. Customers may move from one segment to another and segment definitions may evolve.

'Customer Relationships and Channels' describes customer treatment and reciprocity and the kind of outlet used. Changes are emerging due to the digital revolution, for example user communities are gaining in importance. On the one hand, automation is taking over from telephone conversations, while paradoxically, on the other hand, customers become more involved in co-creation and customisation. The latter phenomenon in particular deserves extra attention in the context of the circular economy.

Channels then come naturally as the ways in which your company communicates with customers. The canvas publication mentions sales force, web sales, own stores, partner stores and wholesalers as channel types. The phases are awareness, evaluation, purchase, delivery and after sales. This interpretation breathes the air of the 'sell more sell faster' model. Circular business would require a much greater emphasis on two-way communication. The core medium is of course your product/ service itself; a point that is often overlooked.

'Revenue Streams' underscores the customer part of the canvas. There are two basic types: the transaction income from one-off payments and recurring income streams, the latter being more in line with the circular concept since it includes time as part of the value proposition.

And here we have it: The Value Proposition. It is the core of any business model and in a circular business model can easily become the most complicated building block, for here value is no longer a one-off consideration. You could say there is a line-up of propositions, or maybe even a proposition that is continuously being adapted according to circumstances.

The proposition is the outcome of the composition of the whole canvas. When time explicitly forms part of a canvas composition, you could arrive at a business model movie in which the elements are continuously fine-tuned. Instead of designing the business model, a new activity emerges: business modelling, which is a form of planning. Here we can quote former US president Dwight David Eisenhower: 'Preparing for battle I have always found that plans are useless, but planning is indispensable.' Replace 'planning' with 'business modelling' and turn circular.

Round and round they go without ever wearing out. Rolls Royce Phantom, Märklin model trains and Thorens turntables.

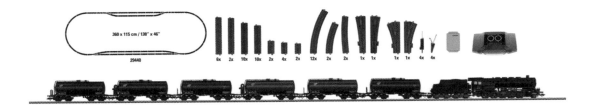

Talk to anyone about products that last and immediately familiar examples will pop up in the conversation that fit into the 'Classic Long-Life Model'. Making high-quality and long-lasting products at the right price is what people tend to regard as the one and only option. It is, indeed, the obvious one to start with.

THE CLASSIC LONG LIFE MODEL

The Classic Long-Life business model is very well represented. As a matter of fact, there is no commercial field without classic products with a reputation for endurance. If you look at them a bit more closely, however, they all lead different lives, and not all of them equally comply with the principles of this book. Probably the largest product is the Boeing 747 Jumbo Jet. Passenger carriers function very intensively over an operational period of between 30 and 40 years. The 747 long-life perspective is of course a business to business affair, which is inevitable since aircraft require considerable investments and careful maintenance. Passengers benefit from the international travel service offered, but the plane is not a product for them to exploit. They just sit in it and wait.

Some cars have a reputation for endurance. Everybody knows Rolls Royce for its traditional chic sturdiness. On the other hand the Fiat 500, the Volkswagen Beetle and the Mini are classics, but their long life mainly concerns their image; few originals are left. The majority of all the Porsches ever produced, on the other hand, are still on the road. Apparently the word 'classic' alone does not cover it all. Long life is what counts. Next, we must determine what it is that has to live long, in order to establish the value proposition. The answer is: individual products. A long product lifespan concerns first and foremost the product that users buy, such as Lego, a Herman Miller Aeron chair, or perhaps a Chanel 2.55 handbag.

Not every product is suitable for a long lifespan. It was not all that long ago that a father gave his son the watch that he was given by his father, who in turn got it from his father. The idea of a similar tradition around a smartphone would be silly. Technology develops at such speed, that even the most luxurious flashy diamond bling version is outdated even before its first text message arrives. The luxury watch paradox is that such an accessory may fit in the world of the Classic Long-Life Model, but that nowadays the possession of a superb functional timekeeper is rather pointless.

Running a business producing long-lasting products is normal practice for the German private company Miele. Their household devices are tested to guarantee a functional lifespan of 20 years (which is different from being designed to last 20 years). This implies that they cannot follow all market vagaries. Currently, vacuum cleaners are equipped with ever longer cables so as to be able to cover a lot of floor area without the need to move the plug to another socket. But those cables are flat (to save space) and often get twisted. Miele's round ones are shorter, but better in the long run.

A long lifespan requires service and repair. For that reason, Miele runs its own servicing organisation. They aim to be the most reliable high-quality company on the market.

For luggage company Eastpak, the perspective is different. They belong to VF Corporation which also supplies other brands, such as North Face and Timberland. Eastpak guarantees their products for thirty years. Nevertheless, in order to be credible in the market in which they operate, their products also have some fashion appeal. Eastpak does repair products, and occasionally they receive a note accompanying a product submitted for repair, reading: 'Even if you cannot repair it, please return it'; a true sign that the bag has become part of someone's life. Eastpak is looking into the possibilities of refurbishing or remanufacturing, but considers this an unlikely option. Weighed against the price of their products, recycling is more obvious.

VITSOE 'You invest in Vitsoe for life. Vitsoe's customers tend to take it with them when they move house, adding pieces as they go. Which means there's another decision you never have to make – getting rid of it. People just don't.' Hugh Pearman, journalist, The Sunday Times

Designed by Dieter Rams in 1960 and produced continuously ever since, the 606 Universal Shelving System is a carefully conceived kit of parts. At the core of the system is the aluminium E-Track that is attached directly to the wall. Shelves, cabinets and tables are hung from the E-Track by slipping aluminium pins into place. The shelving system is produced in a limited range of colours and materials. This helps to ensure supply for the future.

Vitsoe's intent has been to avoid built-in obsolescence by making furniture that is discreet, adaptable and timeless. There is no pandering to fashion. Vitsoe encourages customers to start by buying less furniture so that they can add, rearrange and repair when needed. Importantly, customers are urged to take their furniture with them when they move and thereby ensure continual reuse.

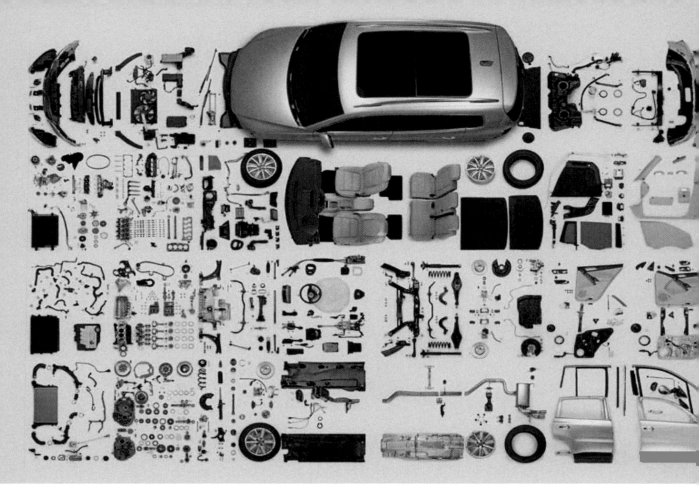

In Europe, the automotive industry is struggling with decreasing sales. In a flat market, car manufacturers can only be successful by improving the perceived value of a car at the point of sale, or by increasing the total revenues generated throughout a car's lifetime. As it turns out, the after-sales market (repair services and sales of spare parts and ancillary products) is highly profitable.

BUSINESS MODEL #2

PRODUCT COMPONENT OF VALUE PROPOSITION
SERVICE COMPONENT OF VALUE PROPOSITION

When a long-lasting product is completely dependent upon a replaceable part with a limited functional lifespan, then the business model is called 'hybrid'. The hybrid value proposition is based on a combination of at least two products that are useless when not combined.

THE HYBRID MODEL

It is interesting to have a brief look at the history of this model, which now has a name, just to explore some business opportunities. Small requisites, like ballpoints and razor blades, were early propositions. At present, ballpoints are so cheap and omnipresent, that refills are only purchased for luxury pens. Razors, however, are a true money-maker. Grips are not designed to be high-profile and long-lasting - although exquisite versions do exist - but they perform satisfactorily. Music records too are early hybrid modules. Via cassettes and CDs they have now evolved into nonmaterial digital media, with films and images and games. They have so little value, and can be accessed in so many ways, that hybridism has virtually disappeared.

At present, the Hybrid Model has proven value for the shaving systems, printers and coffee, and also for some other devices such as air cleaners. In this business model, the indispensable exchangeable part is not just a supply of ink, or taste, or sharpness. It needs to facilitate high-quality functionality, as defined by the supplier. On the other hand, there must be a clear limit to the lifetime of the exchangeable module. At a certain point it is empty - an inkjet cartridge - full - a vacuum cleaner bag - worn out - a razor blade - or just plain dirty. When that is the case it tends to block the system in such a way that for the user it is easy to see, and better yet to predict, when replacement of the element is due.

So in this model the long-lasting product has, as it were, outsourced its wear and tear to one or more parts that can easily be replaced, at a certain price of course.

Apparently the value proposition of the Hybrid Model is determined by a careful balance, partly achieved by providing the main long-lasting product at a low price: a low threshold to get the model going. The user should be prepared to regularly pay for a refill. This means that it should convincingly guarantee perfection and not be too costly when compared with refills offered

by competitors. The word 'convincingly' is important here. For existing printers, suppliers usually resort to warnings against using alien cartridges. Warnings are tricky - they may be counterproductive. The value proposition should embody a feeling of enrichment.

Users like to be in control. This aligns wonderfully with cost reduction through mass production. For printers, four separate cartridges for every printing colour may constitute a better offer than one for all colours. Something similar is imaginable for coffee: compose your own flavour by combining cups, pads or whatever. Judging by the way the hybrid system is developing, it could become valuable for more food products, perhaps with 3D dish printing.

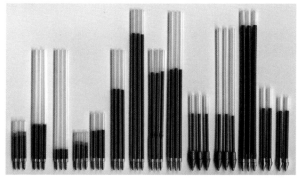

Replacement parts have variable lifespans as well

A totally different option is batteries. Theoretically, they could belong to the hybrid module species. They live in a foggy universe: when are they empty? Some can be recharged, as part of a product or separately, others are preferred as products, with their very own brands. An attempt to introduce exchangeable batteries for electric cars failed. There may be two reasons for this failure, at this stage of development. The first is that drivers are not inclined to accept foreign batteries in their car; the second is that a battery simply provides power with no specific controlled features, such as colour or taste.

2018 Nintendo Switch game - Super Smash Bros. Ultimate sold 5 million units within its first week, making it the fastest-selling Nintendo game of all times. It also has an all-time high record for launch-month dollars sales for a console exclusive: 12 million units sold in 3 weeks.

GAMEBOY Today's computer game consoles are extremely powerful 3D workstations disguised in diminutive plastic enclosures that belie their raw number crunching powers. The gaming hardware is sold at or slightly below true manufacturing cost level because the real money is to be made with the gaming software.

And make no mistake about the business potential of this Hybrid model, although it may be declining somewhat. So far the all time record is held by the fourteenth instalment of Grand Theft Auto for Xbox 360 and PlayStation 3 which did 12 million units, bringing in 800 million dollars in worldwide sales in its first 24 hours. That made 17 September 2013 the biggest launch day ever, for any piece of entertainment, any movie, any record, anything at all. In the next gaming episode the total number of players are likely to set different records. In 2018 EPIC's 'Fortnite' had a weekend with 3.4 million gamers playing online simultaneously.

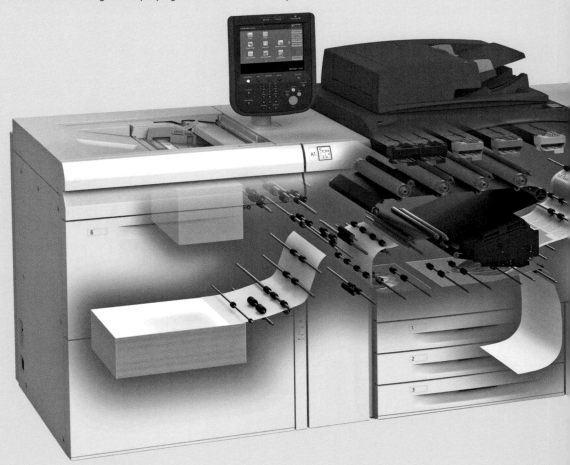

BUSINESS MODELS EVOLVE Herman Miller's Design for Environment (DFE) policy thoroughly evaluates products. In the design process four questions are used: Are the materials homogenous? Are common tools used to separate them? Did it take longer than 30 seconds to reverse a connection? And have the components been marked with their material type? Each and every component is scrutinised based on these factors and assigned ratings on a spreadsheet score card, allowing the design team to accurately evaluate the results.

The focus is on maintaining high standards while incorporating increasingly more environmentally sustainable materials, features, and manufacturing processes into new products. The Design for the Environment team has initiated a protocol to guide this effort. One of the main design tenets is durability. They design for repeated use, repair, maintenance, and reassembly, using standard parts as often as possible. During the new product design process, the team meets with the designers and engineers to review the whole product as well as incoming packaging, and potential waste generation.

CARTRIDGES AND SERVICE Toner can only make its lasting impression on paper with the help of a printer. Companies are cost conscious and well aware that in a hybrid business model the supplies and not the actual printers are the primary source of revenue. Xerox Versant 80 Press has a toner cartridge system and a modular design concept as well, to facilitate flexibility.

In some printer or copier instances a business gap emerges because affordable remanufacturing is becoming a new part of the trade. LMI in Phoenix Arizona does that, and not just cartridges, but everything. They offer refurbished, but also remanufactured printing machines.

S(ch)austall, German wordplay with pigsty and showroom. The gap exploiter recognizes opportunities that others overlook. Like the adaptive re-use of a 1780's pigsty, transformed into a showroom by FNP Architekten from Stuttgart. Or the sales of vintage vinyl records. Once considered an outdated technology, but on the rebound in an immaterial MP3 world with musical concepts that manage to bridge gaps across generations.

For this model, it is important to be aware of the fact that the specific 'gap' to create a market needs to involve an existing product, and that the exploitation concerns leftover value and lifespan. Entrepreneurs are always looking for gaps in markets, to muscle in and make money. They have to, for obvious reasons. So the Gap Exploiter Model concerns only a limited selection of gaps. However, this does not mean that only a few exist.

THE GAP EXPLOITER MODEL

First of all, they exist at every economic level. The less affluent a society, the more people are forced to make a frugal living by squeezing small amounts of money from collecting and selling plastic bags, or other almost valueless waste. The skilled are slightly better off. To repair scooters they run a 'garage', consisting of a plank with some tools placed against a tree along the street. But in rich regions there are maintenance companies for very prestigious cars. Maintenance chains exist as well, for standard services like tyre replacement or air-conditioning checks. Their workshops are cheaper than the ones operated by the original brand and they are often willing to use cheaper third-party parts. The internet has become 'part-searching heaven'.

Most gap exploitation, however, consists of services. Many professions are defined by repair and maintenance: from watchmaker and piano tuner to upholsterer, and computer, bicycle, washing machine and central heating repair man.

Arrow Value Recovery is a company that picks up depreciated computer systems and mobile equipment from companies and wipes out sensitive data. Sometimes the equipment they collect turns out to be unused, in the box it came in. To create revenue, they check and refurbish machines and sell them to other companies; they provide a clean system. In addition, they sell computers through a dealer network, handling a mix of new and used machines. They say that it is only now that companies start to realize the importance of a return service for their equipment. The risk of information leakage and the fact that old systems represent value makes clients understand that returning your digital systems to safe hands is just as important as investing in new systems.

Recognition of re-activities needs some effort. The company Tedrive manufactures stainless steel steering units, engineered in such a way that they can be produced without the expensive moulds needed for more common aluminium ones. This flexibility enables Tedrive to cost-effectively produce units to fit in different types of cars, each with different angles between the steering column and the horizontal plane. This is unique, and atypically allows them to even cater for production of small series, which is usually an expensive route in the realm of the automotive trade. In addition, they remanufacture selected damaged units to a quality level that is literally as good as new. These have to be marked 'R', which is the catch. Remanufactured and new products cannot be mixed, even when of identical quality. This is not just an issue between company and client: laws demand this clarity of status. This to a certain extent limits the opportunity to add value. It is a small print detail of the traditional preference for 'new' over 'quality as required'.

Two fields of gap exploitation have not yet been mentioned. The first of these concerns unsold goods. It can happen that a company produces more items than it can sell. The result is 'dead stock'. Sometimes these products are kept for a certain period of time just in case they are needed; for example windshields, which were applied as shelves in a shoe store by 2012 Architects in one of their designs. Dead stock can also consist of products that people no longer want. Droog Design in Amsterdam runs an interesting project entitled 'UP', in which they invite designers to add 'design' value to such products. This is now happening on a small scale, but it is a principle to keep in mind. The same is true for individual designers who turn old and worn products, from car tyres to vinyl records, into different new objects: a swing, a fruit bowl etc. Value apparently also depends on how you look at things.

REACTIVE AND PREVENTIVE MAINTENANCE

There are two maintenance principles. Reactive maintenance comes down to taking measures after something has gone wrong, too late really. Preventive maintenance on the other hand, focuses on identifying potential trouble spots and items, and act on a special schedule, well before there is an issue.

The former can lead to unexpected costs for parts and labour, plus it can mean lost time and money because of a temporary shutdown, with all due consequences.

Upfront costs from prevention by contrast are predictable. They are based on knowledge, experience and reasonably precise estimations of failure probability. The advantage of preventive maintenance is in fact part of the observation that it is far more important to regularly peek into future occurrences than is commonly thought.

REPAIR ON WHEELS On the streets of Dili, the East Timorese capital, a group of young boys run a bicycle repair shop. Forty percent of the population lives on less than $1 a day, so every extra penny is vital. From East to West gap-exploiters take their chances.

Park your bicycle, travel by train and rent a bike, all combined in the train payment system. The original yellow and blue OV (Public Transportation) bikes usually lasts four years, before being written off. They're not being demolished. NS Railroad Services take them to Roetz-Bikes for a complete makeover. Roetz takes them apart and checks what can still be used. About 70 percent of the original bike is salvaged to continue functioning.

BUSINESS MODEL #4

PRODUCT COMPONENT OF VALUE PROPOSITION
SERVICE COMPONENT OF VALUE PROPOSITION

Access is similar to short-term ownership. It is a step taken when full-time possession of a product is unaffordable and/or unnecessary. Access to a product is complementary to sharing it.

THE ACCESS MODEL

Sharing is currently a buzzword in designer circles, due to the rise of social media. It has a utopian ring to it. Where living standards are low, sharing may well be the only way to afford commodities. Villages where there is only one PC do still exist. Increasing affluence turns this social necessity into a kind of deal. In the beginning of the 20th century, when functionalism made its mark in architecture, there were apartment buildings with shared cooking and washing facilities; sometimes you had to wait your turn to use something. After the Second World War economies grew quickly and now a cooker, a microwave and a washing machine are considered to be basic needs.

Thinking in reverse, selling shared use is also a way to make big money. This phenomenon occurs in the timesharing business, where people get a mortgage on a house or an apartment for a limited time, say two weeks, per year. In between these considerations we have the pragmatism of rental services, mostly of tools and equipment. Sharing, as a phenomenon, benefits from developments in digital communication. In some areas, websites offer the opportunity to check whether your neighbour happens to own the ladder you need. However, a true commercial sharing option is rare. Greenwheels and Car2Go are viable value propositions for sharing cars. At the moment, this is still a niche, but it is one that may grow. Together with the positive attention for electric cars, sharing is favoured by the realisation that automobiles are primarily a form of transport. Some cities at present provide systems for bike sharing, an idea that previously failed in Amsterdam in the 1960s, but is clearly getting a second chance all over the world, from New York to Bangalore.

Of course, like all business model types in this book, this one also needs to address the economic circle. Offering access to a product is not incidental. It belongs to a family of value offering opportunities, such as advice, or perhaps maintenance of the client's own tools.

From the consumer point of view, the decision to temporarily use a product supplied by a provider rests on a balance between five considerations.

The first is relative affordability. If renting is barely cheaper than buying, chances are that products - for example wine glasses - will be bought rather than rented. Apart from that, the product offered must be attractive and well maintained. The second consideration is a perception of freedom. Many people only own an electric drill to be able to hang a painting in the bedroom whenever they want to. People own much more than they use which is why it is so difficult to compete with economic material growth. The third consideration is identity or, as it used to be called, status. Most possessions serve as self-confirmation rather than being meant to function. People buy to show who they are. Sharing and renting somehow have to find their way in. The fourth is time, expressed in duration, or number of times. This concerns straightforward functionality. If a client needs a sound system, a tile cutter, a trailer or a mobile crane just once (for the time being), or an afternoon, or three weeks in the case of a recreational vehicle, he may be inclined to share and rent.

There is one final issue that needs mentioning, which is accessibility itself. It is linked to the second consideration of perceived freedom. For clients it should be easy to get what they need. A good example is the bike rental service provided by the Dutch public transport system. The same card can be used for train trips and bike rides. No questions asked; payment automated: that is access for you.

Public bicycles in Amsterdam 1966 and Paris 2014

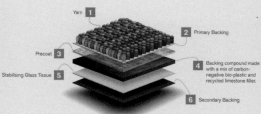

Yarn **1**

Primary Backing **2**

Precoat **3**

Backing compound made with a mix of carbon-negative bio-plastic and recycled limestone filler. **4**

Stabilising Glass Tissue **5**

Secondary Backing **6**

INTERFACE CARPET TILES Interface® is an early Access Model adopter, a well-known example by now. As early as 1973, the company pioneered selling the function of floor covering with a carpet tile system. Interface® aims to be the first company without a negative – and possibly with a restorative - impact on the environment by 2020. They replace worn down tiles, maintaining flooring quality. Retrieved tiles are recycled. The company is in control of the entire material cycle.

A circular system doesn't exclude development. CircuitBac Green is the latest backing innovation. It contains a mix of bioplastics and mineral filler. The biobased content provides a carbon-negative alternative to existing backing materials, meaning it absorbs more carbon than it emits during its production

PERFORMANCE IS EXPERIENCE *For his 'Weather Project in Tate Modern in 2003 Olafur Elisasson placed a giant 'sun' consisting of hundreds of lamps in the former turbine hall. Rolls Royce Lease finances aircraft engines in their engine maintenance programs. Commercial aircraft operators rent engines to ensure that aircraft can remain operational without loss of revenue when engines have to be removed for maintenance or repair.*

BUSINESS
MODEL
#5

PRODUCT COMPONENT OF VALUE PROPOSITION
SERVICE COMPONENT OF VALUE PROPOSITION

There is a very good example of performance. It does not concern any particular product, but superior performance it is. In Mumbai, India there is a system where moderately paid delivery persons called Dabbawalla pick up hot lunches from the homes of office workers, deliver them to the right place, and return the empty boxes afterwards. Each and every weekday about 4500 people deliver up to 200,000 lunches faultlessly, on time. Statisticians claim it is impossible, but it just happens.

THE PERFORMANCE MODEL

That is the essence of performance. It works well for the client and, simultaneously, it goes unnoticed. Products can mediate that kind of service. The value proposition concerns not a certain product, but functionality. Something is taken care of for the client, who pays a certain fee in return. The first prerequisite is that the provider owns and maintains the product, the advantage being that quality responsibility rests with him. Therefore the provider benefits from product endurance. Moreover it is profitable to handle the whole product value circle. Maintenance, repair, upgrading, second hand parts traffic, and eventually recycling: in brief, all material flows are in the producer's hands.

This is the reason that some even advocate the end of private ownership and suggest that value propositions should always be limited to function. This idea overlooks two points: consumption fulfils more needs than technical functionality alone, and owning something is a stronger incentive to look after it than just using it. Of course we also have the other four business model archetypes, each with different possibilities to run the circle. Since business models in this book revolve around products and ways to prolong their lifespan, we will explore what kind of product service mixtures may qualify for the Performance Model.

The most obvious group consists of products that are in use for what they do, rather than for what they are.

The symbolic value of, for example, copying and printing machines is negligible. Computers and television sets also qualify. Although they may represent some status, what appears on their screens is what counts. More distant and abstract fulfillers of function that are no longer really classed as products, data clouds, clearly

fit in the performance way of thinking. Transport is also supplied as a function; London's cabs, for example, run on the basis of a performance based business model. However, if the cab should be submitted to private ownership (sold) after a certain period of time, the business model archetype shifts to the Classic Long-Life Model.

There is a domain of functionality that may be open to the Performance Model. It deals with aspects of conditions in a home. Interior climate is an example. Experience with the emphasis on performance, however, is limited. At present, consumers can lease equipment as a financial arrangement. In future it may become feasible to make a shift towards performance for other experiences that concern the home; for example, if one were to translate comfortable furniture into 'a contemporary lounging experience'. The point is that function does not depend on specific products.

Paying per amount of light, is futile for a reading lamp next to a lounge chair, but the advantage of Pay per Lux systems (Thomas Rau, Philips) becomes huge when the scale reaches buildings, or neighbourhoods, or even cities.

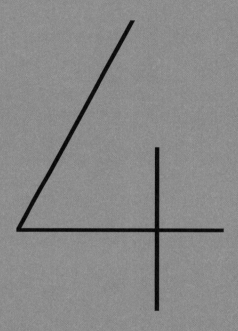

SHARING THE FUTURE

NO MORE URBAN BIKE SPRAWL *Renting your 'own' bike may become more successful than borrowing systems. Swapfiets (fiets is Dutch for bike) guarantees an always working bicycle, with a blue front tyre, for a modest monthly fee. One is offered a diverse range of models and prices and if something breaks the bicycle is repaired or replaced within a day. The client experiences a feeling of ownership, and is likely to handle the bike more carefully. The company started in the Netherlands and is growing rapidly.*

ANCIENT TREES IN OXFORD *The short and simple story – or perhaps, legend – of the oak beams at New College in Oxford epitomises the tremendous value we can reap from long-term thinking. It was said that when (then very) New College received its Charter in 1379 AD, a grove of oak trees was planted on the first bequeathed plot of land. These were destined to replace the roof beams of the Hall, since it was known that on the long run oak wood always becomes infested with beetles. This plan was passed down from College Forester to College Forester for six hundred years. 'You don't cut them oak trees. Them's for the roof of New College Hall!'*

True dark forests of orphaned bicycles scar the fringes of Chinese cities. They illustrate the common carelessness with which products are released into our daily life. The bikes are meant for cheap sharing and convenience, but nobody seems to have taken the time to investigate the need for a bicycle rental system, let alone what the logistic effects of sprawl are, or how to deal with maintenance and repair. Nobody feels responsible for what happens after delivery deadline, which really is the start of the product lifeline.

FAR BEYOND DEADLINES

The emphasis in current economic practice is on quick delivery. The awareness of what should happen next, is low. Online trade, where clients often return their purchases in a clearly used state, get their money back from companies that feel they have to be both lenient and forced to discard returned goods to avoid their brand being blemished. This is just what happens during the brief episode of sales. When the purchase has been excepted, generally negligence strikes

Among most producers and retailers, preparation for after sales activities to sustain the value of what they have sold is minimal, if not absent. There are some exceptions, for instance in the car industry, where repair and maintenance are common. Refurbishing office furniture is not all that exceptional and purchasing refurbished smart phones is not unusual. There even are a few patches in the world of fashion, where trade in used apparel is increasing.

The fact that some industries do think ahead – and may have done so for a long time already - implies that principles are available, but that an accessible systematic approach is only just starting to be developed. In general managing lifelines consists of research, exploration and implementation. Research is finding answers to questions on the state of trade:

How are we doing currently with our products in the user sphere? Do we get many returns? What precisely is the matter with those? Who act as repairers, cleaners and second-hand dealers? How many of our used products can be purchased online? And which ones?

Buying back some products and interview former users can provide a lot of information that may be much more useful than your standard survey.

Exploration starts with an assessment of the lifecycle of the kind of products to which your product belongs. E-bikes are on their way up, coffee tables don't have a clear peak, laptop computers have reached maturity and I-pods are extinct, to name some examples.

Future developments are important to establish too. It helps if they are divided into layers of predictability. For instance, it is fairly certain that electronic technology will become cheaper and smaller, and that human life expectancy is on the increase, resulting in a rising number of elderly people. We also know that the atmosphere and the oceans are getting warmer. The emerging effects of these changes, the second layer, are far more complicated to foresee, because we interpret and influence them. Take for instance, the effects of social media: very few, if anyone, saw them coming. There may be similar human phenomena on the verge of ambushing us, but they may also generate new opportunities. It is important to be on the lookout.

When former British Foreign Secretary Harold Macmillan was asked what he feared most as a politician, he replied: 'Events my dear boy, events'. That is the unpredictable third layer of future (there can be no definite article before future). It can be anything from meteorite impact to an assassination. One can prepare for disasters or lucky strikes, but without certainty that measures make sense.

Wide range future exploration should not be a once and for all effort. Circumstances change, and this is true for our perception of future as well. That is why checking and adapting a broad view on developments to come should be a regular business and design activity, which should be consequential too. Exploration is active and provides reasons to extend networks and embark on new activities, to be able to sustain product value. Landscape design is an excellent example of taking future into account: conditions for growth and development are created for plants to grow and be maintained.

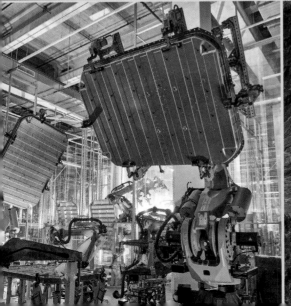

On the hit list
Not yet endangered
Inert
Radioactive

ENDANGERED ELEMENTS Satellite data communication and navigation services and devices are taken for granted. These space invaders can also locate the very materials they need to function.

Computational clouds exist in power hungry climate controlled spaces on data storage devices containing endangered elements, in facilities like this Microsoft specimen. The orange elements in the periodic table above are dissipating into products.

Electric car manufacturer Audi and Umicore together developed an efficient battery recycling system. Up to 95 percent of materials is recovered, including metals such as cobalt, nickel and copper. Reusing battery materials reduces CO_2 emission from manufacturing, but it may also reduce cost. Currently, difficulty is in retrieving used batteries.

SMART BONDS *When Richard Gurley Drew joined 3M in St. Paul, Minnesota in 1921, it was a modest manufacturer of sandpaper, which is a process of making sand stick to glue coated paper sheets. While testing their new 'Wetordry' sandpaper in auto workshops, Drew was intrigued to learn that the two-tone auto paint jobs (popular in the Roaring Twenties) were difficult to manage at the borderline where the two colours met. They initially started experimenting with 3M sheets of glue coated paper without the sand for masking jobs, but this left unwanted traces of glue on the paintwork.*

After two years of experimentation in 3M's labs, Drew invented the first masking tape (1925), a two-inch-wide tan paper strip backed with a light, pressure-sensitive adhesive. In 1930 he came up with the world's first transparent cellophane adhesive tape (called sellotape in the UK and Scotch tape in the United States). During the Great Depression, people began using Scotch tape to repair items rather than replace them. This was the beginning of 3M's diversification into many different markets. It helped them to flourish, in spite of the Great Depression. The company started developing and producing all sorts of tape products, including magnetic recording tape.

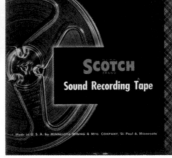

Although designers can tempt entrepreneurs to reframe their markets, for instance by including refurbishment, in practice, those who commission designers operate according to a business model and decide on new requirements to apply in a new product service combination. Since, more often than not, the business model is based on the *Sell More, Sell Faster* principle, this whole theme of mutual influence is hardly being challenged. Since including the management of future developments demands a wider view on business design, it's about time.

COMMON VISION

Taken one by one, each business model has a direct effect on the formulation of design requirements. 'Classic long life' is almost a requirement in itself. To some well-reputed companies, it is almost literally a slogan to sell products. All it needs is subdivision and specification. Technically, it implies that parts must have a low failure probability and that maintenance, repair, upgrading, etcetera, all relevant future occurrences, need to be dealt with properly. The appearance needs to be what is usually circumscribed as 'aesthetically appealing', but 'timeless'. Moreover, certain target groups need to be able to identify with it.

The hybrid business model causes subdivision of requirements. The main product part is subject to very similar characteristics as the previous kind, but there are one or more exchangeable dedicated filling elements. It can be anything from a simple container for detergent or coffee, to a mechanically complicated distributer of toner for a printer.

This is a category that is defined by material flow (Products that Flow), which implies that responsibility for handling disposable elements, which can range from recycling leftover materials, all the way to advanced part remanufacturing, needs to be clearly defined. Material fall-out needs to be kept to a minimum, through well-organized reverse logistics. Responsibility for proper end-of-life collection and treatment, in its turn can be the user's, but it can also be government, or the original producer, or a different company, which would define this business as a gap exploiter.

Foreseeing what gap exploitation implies, particularly for design requirements, is a blurry consideration, since it can involve unexpected business development by unknown traders. You cannot design *for* a gap exploiter, but you can design a gap exploiter. Learning from the past is the first activity that should be standardised. The best option is to carefully investigate what happened to your earlier products and to the ones made by competitors, and beyond that to entirely different markets. It makes a lot of sense to broaden horizons, particularly when it concerns unforeseen gaps. Dishwashers are interestingly similar to motorbikes in certain respects. As a matter of fact, all things produced have common properties and necessities.

An extreme example is the porous material that Ferrari needed for aerodynamic control. It happened to be available in the felt pen industry. There may be previously unconsidered partners out there, to share responsibility with. They may be able to point out new requirements to you that help them be better repair experts, or part adapters, or recyclers.

In both the Access and the Performance models the product exploiter retains product ownership. He therefore benefits from sustainable value. This entails that products must allow more intensive particular use and be totally suitable for maintenance and repair. Rental bikes for instance are not handled by users as carefully as they would ride and park their own. They may be not as comfortable as personal human powered two-wheelers, but they are enduring and often built to fit a storage system. Recognisability is quite important. They have to stand out when prospective renters are looking for one.

From the designers point of view – design strategies are discussed in the final chapter – the message to their clients or employers is not complicated, but it can be quite far reaching. All strategies add up to enhancing the ability to define responsibilities and plan action to keep products going as long as is wise.

WHICH PRODUCT DESIGN STRATEGIES APPLY TO YOUR PRODUCT?

Boomerangs fly and exemplify the earliest heavier-than-air man-made flying capability. They have two or more wings arranged in such a way that the spinning from a skilled fling creates unbalanced aerodynamic forces. They cause the object to travel in an elliptical path back to its point of origin.

Most designers are not aware of the fact that everything they do is defined by an implicit business model, and that this is virtually always a model to sell products and then ignore them. Even newly graduated designers who have started their own small scale production of handbags or vases, sometimes find out the hard way that they have produced faulty products and that they have to find time to repair returned items. Little attention is given to this in education, and designers have little practical experience with strategies to design products with a long lifespan.

CIRCULAR DESIGN

Defining design strategies for a long product lifespan starts with the notion that design concerns the whole economic circle of change. The ability to keep on sustaining and adding value is the crux. Therefore, extra requirements for a given new product concern a certain sturdiness and accessibility. This must be added to to all the others, the ones about logistics, brand identity, fire resistance, energy consumption, and so forth. The combination always depends on the project at hand.

Since design for longevity involves reduction of environmental effects, it is interesting to look at differences and similarities with earlier descriptions of 'design for sustainability'. So far, the attention in that domain has focused mainly on closing the material gap, going from new to recycling and onwards, back to (almost) new again. In the case of production, this has led to a combination of requirements with respect to efficiency and clean processing. During their period of use, sustainable products should consume a minimum amount of energy, and preferably produce some. Their main features are easy disassembly and the use of one kind of material for each part, to facilitate shredding and regeneration. There is nothing wrong with these requirements, except that they lack the perspective of maintaining value over time: starting with a strong proposition and nourishing it in order to keep it alive.

The main difference between 'design for sustainability' and 'design for a long product lifespan' is the principle of sturdiness. The quotation marks serve to emphasise that the difference should be cancelled out and that what is now known as 'sustainability' should assimilate lifespan extension strategies.

Currently six strategies have been defined. Their order is based on a concept called 'Product Integrity', a measure for how true a product remains to its original, factory fresh, configuration. When a product leaves the factory, its product integrity index is 1. Disassembly into parts pushes the index towards zero. After shredding

or smelting, when no trace of the original geometry remains, the index becomes 0. The product integrity index is rooted in the assumption that every change made to a product requires energy and raw materials. Different strategies can work perfectly well in certain combinations.

- **1. DESIGN FOR ATTACHMENT AND TRUST** is a holy grail for many designers. They explore the way in which users develop a certain bond with the objects they use. The complexity of attachment and trust is a fantastic challenge.

- **2. DESIGN FOR DURABILITY** is based on defining optimum product reliability. It is a well defined technical field. Ideally a product's durability should match its economic and stylistic lifespan.

- **3. DESIGN FOR STANDARDIZATION AND COMPATIBILITY** is constantly evolving. Digital technology, for instance, has come a long way. There is an interesting field of tension between setting standards and the reality of personal customization.

- **4. DESIGN FOR EASE OF MAINTENANCE AND REPAIR** is a tough one. Maintenance and repair are currently divided among the original manufacturer, the gap exploiting service provider and users. Users are treated in a rather patronizing way. Repair is not allowed, under penalty of loss of warranty.

- **5. DESIGN FOR ADAPTABILITY AND UPGRADABILITY** implies to incorporate possibilities to change a product. Adaptation to different functions by part exchange is common, upgrading less so. Particularly digital technology develops with so much speed that upgradability is limited. Value propositions would have to be different.

- **6. DESIGN FOR DIS- AND REASSEMBLY** is partly new. Easy disassembly is a classic requirement for sustainability. The possibility to reassemble is similar to the previous three strategies, but may include assembly with components of other products to become something different.

With clocks everywhere, including in our computers and smartphones, we don't need watches. Yet the watch remains a cherished accessory that expresses taste through mechanical hyper precision.

Sam New and Sam after nine and fourteen years of hugging for love and consolation.

The Classic Long-Life Business Model directly implies the need for products that evoke attachment and trust. The challenge regarding Attachment and Trust is that they only partly depend on designer influence. Context and user or owner psychology can only be estimated in terms of probability. Many concepts and proposals have been produced that stimulate the emergence of an emotional bond between user and product.

ATTACHMENT AND TRUST

The trigger to explore this theme was the rising awareness of the fact that digital technology no longer defined shape and functionality in the way that mechanics used to do, leading both to a loss of 'mutual understanding' between person and product and to opportunities to define product identity in metaphors. Interesting student projects took place at Cranbrook Academy (Bloomfield Hills, Michigan, USA) around 1990. One of these involved a small TV set that would humbly withdraw when switched off.

In 2000, Droog Design, together with advertising firm Kessels Kramer, started a project entitled 'do create', involving objects that the user could appropriate through a certain intervention: hit a steel cube to make a chair, smash a rubber vase covered with a ceramic coating. More projects like these followed. Martí Guixé designed a lamp covered with blackened glass entitled 'do scratch'. The user could scratch the surface to make it forever unique. A year later, designers Fiona Raby and Anthony Dunne carried out the Placebo project on the subject of product character, using furniture with surreal features in people's homes to see what would happen. One example was a table with a matrix of compasses which behaved irregularly depending on where it was placed, such as next to a radiator, leading users to believe the table preferred it there.

To investigate the theme of product character, appropriation and attachment, four students conducted interesting projects at Philips Research. These concerned the interaction between user and product, and served to explore the perception of product behaviour. George Tokaya investigated the perception of a small lampshade featuring a sweet and shy kind of behaviour. TU Delft IDE student Bram van Krieken carried out a similar project, in which a water cooker reacted to the user by sneakily turning its handle away when the user

had forgotten to turn it off. Marc Padró focused on lighting, reasoning that an LED bulb is likely to outlast the fixture it is part of. He designed a black lampshade that would gradually become more unique by scratching light leaking patterns in the black lacquer in reaction to surrounding sounds. The difference with Guixé's 'do scratch' is intriguing: scratching by hand as opposed to scratching by system.

There is another side to this matter, which has hardly been investigated. Attachment is not equally important for every product. Use and possession of products are generally defined by a kind of obvious normality. The decision to get rid of something is determined by a glitch in what could be named 'suspension of distrust', an analogy to the term 'suspension of disbelief', that determines the quality of a story. You don't put something away to be forgotten if it doesn't disturb you for some reason: a noise, a dent, a stiff hinge or whatever.

Yet another consideration tends to be overlooked and that is affordability. The question of whether a certain user will buy a new product, will have the old one redone or will accept the unsatisfactory situation is of course also an economic one. The amount of waste depends on conjunctural phenomena, as do decisions to buy things. 'The Queen of Versailles' is a documentary about a very rich family falling back into poverty as a result of the credit crunch. Their enormous house had one large room containing a pile of children's bikes. The mother (Queen), just bought a couple whenever she saw them in a supermarket. Bikes were so affordable that they were turned into junk immediately after purchase.

AFTERLIFE *When a broken AIBO is beyond repair, some parts are salvaged to help revive a fellow AIBO that might still have a few years of companionship left in it. Since the dog is no longer in production, people send in their old or broken ones for this very reason. It's organ donation for toys. So, before the 'medical' procedure, they have a funeral service, for they believe even a robot dog has a soul, and that this should return to its owner, thus allowing the toy to become a mere machine. AIBO is a testament to the success of Sony's engineers in creating an artificial companion that consumers got so incredibly attached to.*

WAVES Sleeping bags full of gaming addicts in front of electronics stores, waiting for the door to open, may fade away. Telmen Dzjind graduated on 'Waves' in which gaming is a product service combination. Gamers don't own consoles anymore, but they each subscribe to a level of needs that they can upgrade with modules according to their increasing experience. Parts no longer needed can be remanufactured and passed on to less experienced gamers. This reduces the amount of required hardware and – this is clever – support attachment between players and what really matters here: the game.

SUSPENSION OF DISTRUST Apps and sounds and content characterize someone's personal smartphone. It is a handheld extension of individual preoccupations. The object as such doesn't really matter all that much, as long as you can find it. You won't replace it as long as you can trust it to sufficiently recharge. In this case attachment addresses the virtual.

BREACH OF TRUST *It becomes increasingly difficult for consumers to ignore the risks of trusting companies with their personal information. It also becomes easier for them to leave brands who break their trust. It really is hard-earned and even harder, if not impossible, to regain once it is lost. When your customers trust your company to take care of their personal data, one misstep can cost you countless customers, negative global attention and sky-high fines.*

DAMAGE DONE Good customer service keeps customers coming back. Bad customer service keeps customers away. It is important that a business has effective and pro-active customer service personnel and a good policy. Keeping customers happy, and getting them to tell others about the great service they received, will propel a business forward, whereas a company's lackluster customer service policy will damage the reputation of the company in several ways.

Some drinking glasses may explode spontaneously. This will not only destroy that glass, but also the reputation of similar glasses in your household and similar glasses in the vicinity of the readers of this book.

Product durability depends on durability of all respective parts. Not all parts have to be equally reliable as long as the ones that wear out can be replaced in time, either preventively or in a repair procedure.

LOW ——————————————————————— **PRODUCT INTEGRITY** ——————→ HIGH

To produce something with a long lifespan is more or less synonymous with designing for durability. The strategy is obvious, but the way to make it work less so, simply because for many products it has hardly ever been an issue: make sure that it works until the client stops caring that it has ceased to be. And let us not forget gifts. Many a label printer, to take a cliché example, is passed on to someone else in the family, or stashed away somewhere, without even having been unpacked. But still, even in this case its quality is important, just in case the next owner turns out to be a user.

DURABILITY

The only way to find out if something works, and how well, and for how long, is to test it. Testing a product requires assumptions about how the product is going to be used, and imagination. A street lamp post, for example, must remain upright with a drunk spinning around it, or with Gene Kelly singing in the rain. On the other hand if a car runs into it, the collision force can be partly neutralised by allowing the post to break. To imagine what could happen in the long lifespan of a given product is the main point of this book.

Probably the most neglected assumption is that a computer represents reality sufficiently well to trust the results of computer-assisted design and testing. Some designers actually think a product can be designed on a flatscreen and then simply be realised. This only applies insofar as it concerns interactive design on a similar screen. Even graphic design requires testing with real ink on real paper. Mechanical functionality is far too real for computers to understand, let alone what will happen to function in the long run. Product behaviour must be tested in practice. Programme models to estimate strength, stiffness and component behaviour during processing provide information, but cannot be trusted without the proverbial proof of the pudding. Testing for durability can take a long time, unless assumptions are made about the product's workload and a concentrated version of the product's life is developed.

Product reliability is measured in terms of the probability that a product can no longer fulfil its function. Products that consist of several parts require multiplication of the respective reliabilities of these parts. Suppose there are three crucial components with a respective reliability of 0.2, 0.6 and 0.4 (20%, 60% and 40% probability that the part will remain intact). This means that the product as a whole will only have a 0.048 (4.8%) chance of surviving. Now consider the reliability of all the parts of a passenger airplane. Isn't it a miracle that it flies?

So far we have discussed technical reliability, but there is also a different kind of reliability, which is quite relevant for the lifespan of products. Reliability perception may well lead to a different technical outcome than actual reliability. This phenomenon emerges when the consequences of failure enter the scenario.

The history of copying machines provides a clear example. Copying machines used to be fairly large: the size of five or six refrigerators. They would function for months on end and then, inevitably: paper jam. An engineer had to come and restore its function, perform maintenance and clean the machine, resulting in it being out of use for two days. When copying machines were further developed and became smaller, from the early 1980s onwards, producers realised that it would be preferable if users could handle defects like a common paper jam themselves. This led to an interesting counterintuitive effect: users were prepared to accept the machine coming to a standstill once a week as long as it took them mere seconds to open the machine and pull out the piece of paper that got stuck. As a matter of fact, the new concept was perceived to be more reliable than its predecessor. Here we note a psychological link between durability and maintenance.

SAFETY AND DURABILITY Before the development of an aircraft approaches the stage of building a flying prototype, all systems are tested together in a complete model called an Iron Bird. It has the same configuration as the airplane will have, except that there is no airframe or skin. In this way mutual influences between for instance climate and flight control can be traced.

Separate structural static tests include: maximum wing bending at limit load, ailerons and spoilers functioning during maximum wing bend, fuselage pressure, and fatigue and flight cycles simulation. Fatigue testing provides data on the aircraft structure response to stress over a long period of time and during different stages of its operation. The A380's fatigue testing, for instance, took 26 months.

A 16-hour flight can be simulated in just 11 minutes. A complete test can take months.

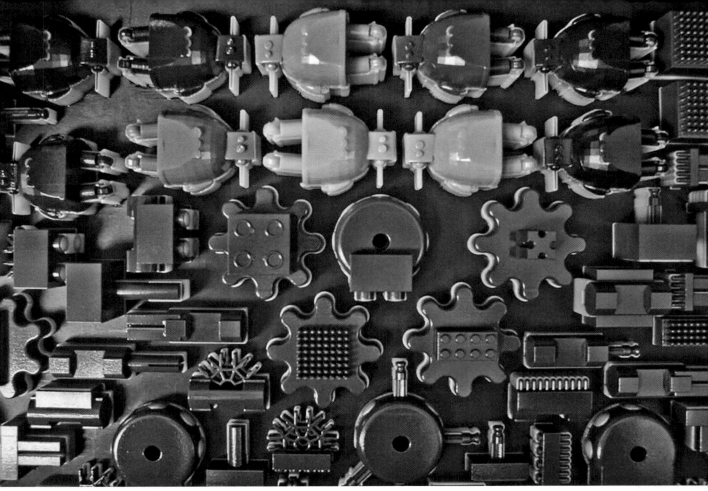

META CONSTRUCTION KIT *Why be satisfied with just one system. With this Free Universal Construction kit one can 3-D print construction blocks that link 10 different kinds of construction toys like Duplo, Fischertechnik, K'nex and Lego.*

ADJUSTABLE LENSES *Joshua Silver, a physicist by background, and his team at the Centre for Vision in the Developing World have designed low cost self- adjustable eye-glasses. The product is aimed at people who need to wear glasses, living in low-income countries where there are few eye specialists and services available to carry out screening or even perform simple tests. These eye glasses can be manufactured at a cost of $19 a pair. The wearer can adjust the glasses to their respective prescriptions by altering dials fitted to the arms of the glasses. These control the amount of silicone fluid in and out of the lenses, thereby adjusting their power.*

LOW ———————————————— PRODUCT INTEGRITY ————————————→ HIGH

Products that exist as monoliths in isolation are a rare breed. In the more common variety they are surrounded by other products, made up of multiple parts. However, when they are connected to other parts, to devices or to whimsical systems with asynchronous lifecycles, they still need to be time-proof.

STANDARDIZATION AND COMPATIBILITY

Think of connections to for example the power network, to bolts, to small plastic attachments, or even to sticky tape that may damage the plastic surface of your product. Design needs to deal with such matters through standardisation, with compatibility in its wake. This is more complicated than it sounds: a Brazilian designer can travel to Beijing and connect his laptop to a video projector over there, which is a miracle.

There is an essay by Douglas Adams, author of The Hitchhiker's Guide to the Galaxy, in which he humorously complains about the omnipresence of what he calls 'dongly things'. They are the connecting cables and electricity adapters that come with many gizmos. Some remain packed in their plastic wrappers, because you do not need them (you had three already). Some of them function, some don't. You never know what kind you need for which gadget. They are the technical equivalent of obesity in the human body, the result of unnecessary extras. If copper thieves would exclusively steal dongly things, they would do little harm and make a fortune. Tediously slow-evolving compatibility forces travellers to buy strange objects that look like puzzles or plastic hedgehogs, just to be able to plug in and charge their laptop. Dongly things have to be conceptualised because of a lack of standardisation.

Standardisation seems to be left in the dark, but it takes quite a bit of effort to acquire. First, you have to know which wheels have already been invented. Not everything needs to be designed. You can buy small parts that do jobs you had not even thought of doing. The next level is to invest in design and equipment to produce your own standard connection. When you are in the front line, you are in a position to set standards. Lego has succeeded in sustaining and expanding their system. A similar concept for electronics called LittleBits has recently appeared on the market. It intelligently supports the creation of electronic functions with colour-coded elements.

Architects can afford to set their own standards. For one large building they can design a unique connection system, usually for a glass façade. The parts have to be produced in fairly large numbers, but cannot be applied in any other building. Nevertheless, they suggest an idea of universality.

Standardisation gets more complicated when a competitor produces a dongly thing that can easily be adapted to your product as well. Behind the scenes specialists and standardisation institutes negotiate for years about precise requirements to make sure one thingy fits to the next little clip.

In digital technology, procedures seem to have eased a bit with the gradual disappearance of connecting cables. Software also requires standardisation. Adapting programmes may be easier than fiddling with wires and hardware, although consumers still live in different universes, depending on the preferred brand of digitalism.

Standardisation on the internet is sometimes also mistaken for an issue. This was the case when it seemed difficult to come to an agreement about standards for typefaces, the use of which depends on copyrights. With the whole world having access to thousands of fonts, it is a potential incentive for nothing less than cyber war. Luckily three typeface designers, Erik van Blokland, Jonathan Kew and Tal Leming, discovered that they could resolve the whole issue through programming. In 2009 they came up with the Web Open Font Format. Apparently, the requirement for a standard is not always self-evident.

Alphabets and other symbol systems are quite successful standards. Together with icons and pictograms they function as an indispensable support to connectivity. Product graphics serve to help users make the right connections. As a designer, you must be fluent in the language of standards.

POWER STANDARDS *Although size, chemical composition, position of terminals and special characteristics of batteries are laid down in international standards, this does not mean that all batteries are created equal. Batteries that are physically interchangeable may still possess markedly different chemical and performance characteristics. In other words, size is not the only thing that matters when replacing batteries. The universal travel plug, resembling a robotic hedgehog, is a solution to the exact opposite problem: tricking a geometric multitude of holes and pins into conducting essentially the same electrical power.*

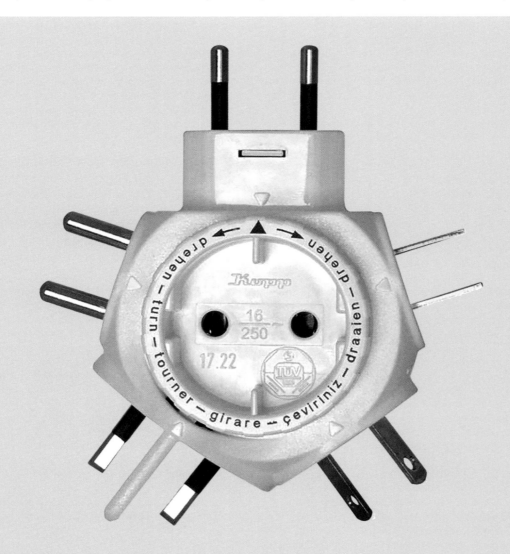

CONSTRUCTION SYSTEMS For one episode of James May's Toy Stories (BBC television) LEGO donated 3.3 million plastic bricks so the presenter and another 2000 volunteers could assemble a full sized LEGO house. This included the bed, working sink and taps, and even a working toilet. The really interesting part of this project is how they figured out how to make it structurally sound. They were dealing with LEGO physics on an entirely new scale. According to May people were really enjoying it simply because these massive piles of coloured Lego were theirs to push together. 'It is a faintly spiritual activity that everybody connects with'.

For the same program he took the question 'How hard can it be to build a motorcycle out of Meccano and drive it around the Isle of Man?' seriously. Meccano is the brand name of a model construction system of perforated metal strips, plates, angle girders, axles, gears and wheels that can be connected by an assortment of miniature nuts and bolts. May and his head engineer Simon Oakley managed to build an entire motorcycle, including a wind-up propulsion system. This is particularly impressive considering that the bike and sidecar had to be able to carry two adults at speeds of 25 kilometers per hour over a rather tricky 60 kilometer course with steep hills and tough turns. They completed the trip, but it shouldn't come as a surprise that it took them three days, mainly because the mechanical drive unit needed to be wound up fairly frequently.

REPAIR OR REPLACEMENT *A piano is not only a part of the art of music, it is also a work of art itself. The machine is extremely complex and has thousands of moving parts. It has a framework and a sound board supporting tremendous string tension, all concealed by a beautiful finished cabinetry. The piano is not as fragile as some other instruments, but it is still subject to deterioration over time. The felt wears, strings break, wooden structures weaken and crack, and the beautiful exterior cabinet loses its finish and elegance. The question always is if the cost of repairs exceeds the price of replacement. Renovating high-quality, large pianos may only cost half of the price to replace the instrument.*

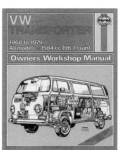

DO IT YOURSELF *The first Haynes Manual (author John Haynes) was published in 1965. It provided detailed instructions on maintenance and repair for the Austin-Healy Sprite. Today Haynes Owners' Workshop Manuals appear in 15 languages. They are widely used both by DIY enthusiasts and professional garage mechanics. Manufacturers generally cooperate with Haynes by providing detailed technical information to evoke customer loyalty.*

DESIGN STRATEGY #4

LOW ——————————— **PRODUCT INTEGRITY** ——————————→ HIGH

Cleaning, brushing and polishing are relatively simple acts of maintenance. Yet even these may put the relationship between supplier and customer under severe stress. Not everything is allowed, and faulty treatment will result in scratches and loss of gloss. Entering the bowels of products is more of a hassle. Stickers warn users not to touch anything and command them to return their (yes, their very own) product to the manufacturer, or, at the very least, take it to a qualified repair person. Product maintenance in a low-income country is quite a different matter, offering unlimited possibilities.

EASE OF MAINTENANCE AND REPAIR

Maintenance and repair are interventions in products within an economic force field, the limits of which largely depend on product design. Even something as straightforward as changing accumulators is cumbersome. A smartphone is really nothing but a battery with a lot of features. Maintenance and repair are labour intensive. In low-income regions anything is repairable, but in high-wage regions maintenance is expensive, and the majority of consumers have sufficient income to replace rather than repair moderately priced products. Blenders are less likely to be opened up to see what is wrong than washing machines. Cars suffer from the severity of their function, but luckily they are also expensive enough to be repaired and to give them their rightfully deserved portions of fresh oil and other messy fluids. Car history shows that technological development has an effect on design for maintenance as well: diagnosis is turned into a simple plug-in system and units that need regular replacement are easy to reach.

This provides an example for the principle according to which other, less expensive products could also be designed. There is a potential for considerable lifespan extension here, by designing maintenance which is worthwhile for all the parties involved. This means that maintenance turnover should generously compensate for a decrease in sales. It should generate money, but also increase reputational value. In addition, a change like this requires a redefinition of the areas that can be opened up and understood, in some cases by users with no professional competence. There could be a role for 3D printing here as well: open up your device, see what's wrong, download a template for a modest fee, print it and swap it for the broken part. Right now, the internet is already a gigantic source of manuals, repair information and instruction videos.

This new viewpoint requires redesign of product interiors, which implies that designers should have access to areas that traditionally belong exclusively to the realm of engineers. Together they can develop new concepts for certain components to be easily exchangeable. These different designs need not be costly; they follow from a different perspective with different requirements, but they do not necessarily involve more effort.

Design for maintenance is the most obvious strategy when the supplier benefits from it himself, which is the case in business models where the value proposition includes supplier ownership.

TU Delft IDE student Maarten van den Berg's graduation project at Philips Research suggests interesting ideas about a lighting system that the company will be leasing to Schiphol Amsterdam Airport, whereby spaces at the airport are properly lit, with the light provider taking care of every aspect, including payment of the electricity bill. This is a typical example of the Performance Business Model.

Van den Berg analysed five existing luminaires, of which only one seemed to be truly accessible for service and maintenance. Important observations were that a modular design helps, but that modules, or screwable attachment of parts, were in themselves not sufficient to accommodate easy replacement. An LED module is lovely, but pointless if you cannot reach it. Consequently he designed a concept for a new 'future-proof' luminaire, allowing for easy replacement of LEDs by more efficient ones. In a broader sense, the designer recommends 'an ecosystem of modules that is likely to be used and produced for a longer period of time'. This is in fact an introduction to the next strategy.

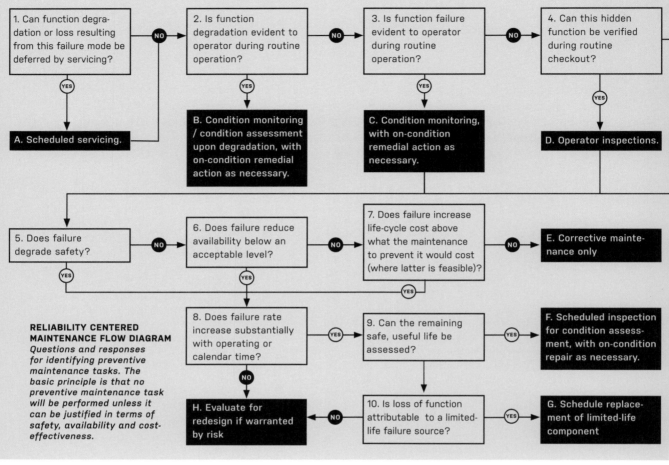

1. Can function degradation or loss resulting from this failure mode be deferred by servicing? — **NO** → 2. Is function degradation evident to operator during routine operation? — **NO** → 3. Is function failure evident to operator during routine operation? — **NO** → 4. Can this hidden function be verified during routine checkout?

1. → **YES** → A. Scheduled servicing.

2. → **YES** → B. Condition monitoring / condition assessment upon degradation, with on-condition remedial action as necessary.

3. → **YES** → C. Condition monitoring, with on-condition remedial action as necessary.

4. → **YES** → D. Operator inspections.

5. Does failure degrade safety? — **NO** → 6. Does failure reduce availability below an acceptable level? — **NO** → 7. Does failure increase life-cycle cost above what the maintenance to prevent it would cost (where latter is feasible)? — **NO** → E. Corrective maintenance only

5. → **YES**

6. → **YES**

7. → **YES**

RELIABILITY CENTERED MAINTENANCE FLOW DIAGRAM
Questions and responses for identifying preventive maintenance tasks. The basic principle is that no preventive maintenance task will be performed unless it can be justified in terms of safety, availability and cost-effectiveness.

8. Does failure rate increase substantially with operating or calendar time? — **YES** → 9. Can the remaining safe, useful life be assessed? — **YES** → F. Scheduled inspection for condition assessment, with on-condition repair as necessary.

8. → **NO** → H. Evaluate for redesign if warranted by risk

9. →

10. Is loss of function attributable to a limited-life failure source? — **NO** → H. Evaluate for redesign if warranted by risk

10. → **YES** → G. Schedule replacement of limited-life component

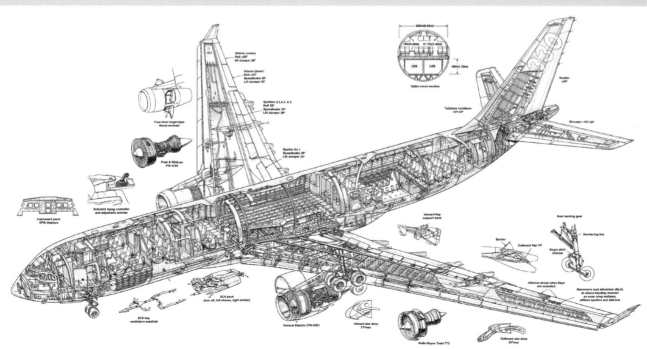

Researchers at the University of Illinois have developed a regenerating polymer that offers a significant leap in technology, so much so that it can recover from a gunshot- sized-hole in just a few seconds.

SELF HEALING MATERIALS Miracles do exist, like a wound afflicted to a living body healing, al by itself. Imagine a future in which smart-phone screens and tires can do the same, heal, just like we do. We would no longer have to replace the things we produce over and over again. The cracked screen doesn't need to go into the trash, because it achieves fresh perfection. It would just need some rest. It would be a world much closer to one without waste.

Researchers from Stanford University created an elastomer, mixed with metal ions, that can stretch up to 100 times its original length and repair itself if punctured. Ions can 'let go' and reattach. They also showed that they could make this new material twitch by exposing it to an electric field, causing it to expand and contract, making it potentially useful as an artificial muscle. This research dovetails with efforts to create artificial skin that might be used to restore some sensory capabilities to people with prosthetic limbs.

As you can imagine, self healing concrete is a total game-changer. It gives us the ability to construct buildings without worrying about damages or intensive maintenance. The first step came from the ancient Romans who mixed volcanic rock with cement, a reason why the Pantheon is still standing tall after over 2000 years. Not only will structures benefit from self healing concrete, also it is a wonderful solution for sidewalks. Smooth pavement can be laid down in cities and suburbs, without having to worry about wear and tear. Self healing concrete was invented by Henk Jonkers, a microbiologist and professor at Delft University of Technology in the Netherlands in 2006. Limestone-producing bacteria will thrive in the high-alkaline conditions of concrete and produce spores that can live up to four years without any food or oxygen. Jonkers finalized his creation by adding calcium lactate to the limestone concrete mixture in order to feed the bacteria so that they can produce limestone for up to 200 years to repair cracks in the concrete.

An option would be to coat a car with this new self-healing material. Scratched or cracked areas would have to be heated to a modest 50 or 60 degrees. This could be accomplished with a blow dryer or with an integrated small electronic heating system. Car paint also heats up when it is exposed to sufficient sunlight. The new coating would not melt, but the crack would disappear.

Nissan is testing a self-cleaning nanotech paint that repels dirt. Nissan describes it as a superhydrophobic and oleophobic paint, meaning it repels water and oils. A Nissan Note subcompact is being tested in Europe with half the car treated and the other half with just normal paint.

FAIRPHONE *is an admirable attempt to develop a smartphone with transparant, accessible technology made from socially responsible mined materials. Repair and upgrading are easy. Parts can be recycled. Its most important quality is that it sets a benchmark.*

Installing system update

SAMENESS *Software seems endlessly adaptable, but variations are always limited to choices made by developers. Consumers are lazy choice makers and often prefer the convenience of default settings. As a result we have considerable variation in sameness.*

LOW ——————— **PRODUCT INTEGRITY** ————————————————→ HIGH

We are really on the threshold of the future right now. Products with a long lifespan are adventurous. They will want to go beyond the time horizon and we will have to prepare them for this, requiring the development of probable product lifespan scenarios. A thorough investigation of the product's 'now' can provide important information.

UPGRADABILITY AND ADAPTABILITY

The future consists of three types of projected facts: things that are known to happen, things that are likely to happen with a certain probability, and downright surprises. The further design ideas are projected into the future, the larger the uncertainty becomes. You may know for certain that in two years' time you will have completed the design of a new electronic kitchen appliance. Digital memory will probably be a lot cheaper by then, but you cannot anticipate that within eight months the food industry will successfully introduce a new insect protein based dinner that could make the sales of your product soar, provided you retroactively make adaptations. That is what adaptability and upgradability are about.

It implies that a design has to include a scenario that is projected well into the future and that is continuously being adjusted, in line with all the relevant changes. To most designers this is a new way of thinking. Scenarios are quite common, but they always serve to imagine the context in which your product will be used, and to explore function. They concern mapping of qualities and features, technicalities and human factors, but they provide no insight into what time can do to a product.

That kind of scenario needs information of a specific kind. The first layer of information needs to be retrieved from the products that are or have been in operation, and from what their users tell you. It is worthwhile attempting to have used products returned. What you will learn - damage occurred, stupid details, material properties and information from previous owners - is well worth paying for. There is an advantage here for companies with Access and

Performance Business Models, since handling returns is a normal process for them. The second layer is general information about technological developments and how they may influence your product in the future. For example, digital smartness is impregnating just about everything. This may imply that the theatre chair you create should, at a certain point, be able to light up with a small blue dot, to indicate that it is empty. Chairs may be linked to the smartphones of the people in the audience. The third layer consists of more general kind of data, about developments in society that may become relevant to your product. You need to be alert.

The scenario of change is bound to have consequences for your design. You can divide it into functional levels, some of which will change very slowly, if at all. These could concern the general structure. Others have to be easily adaptable, by exchanging or adding modules. They can be technical updates, but also parts that provide new functionality because the user has discovered new opportunities or is confronted with new needs due to children, accidents, age, or other radical life changers.

A level of very fast adaptation is the now familiar fully automatic software upgrade. It is almost like a rain shower - unexpected, sometimes terrible, but you get used to it, except of course for the weird 'time to install' estimations it comes with.

PART LIFETIME CATEGORIZATION *Some samples from the bugaboo program: The frame has three use cycles / Seat fabrics and sun canopy to be replaced every use cycle / Wear and tear parts like wheels and straps can be replaced within one use cycle.*

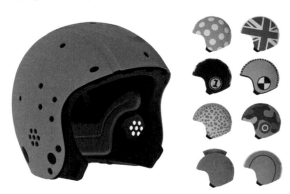

ANT FOR LIFE Life is rough for a stroller. Not only does it have to suffer from the, often undisciplined, behaviour displayed by the babies, or toddlers, or children it has to carry, but also, it is taken for granted by parents who always care for more for their offspring than for their equipment. Strollers just have to live up to expectations. Strollers are the main product of bugaboo, an enterprise that wants to facilitate free and responsible movement for all. It is a very good example of a company that tries to design by circular strategies as best they can. Their latest, the bugaboo Ant, is truly part of a material flow and created to last in almost all respects. It is of course modular, without any riveted or glued attachments. Therefore, it can be easily taken apart and put together again. Parts can be replaced and upgrading to more convenient levels of functionality is optional. Some of the materials, such as the fabrics of the seat and sun canopy, contain a considerable amount of recycled components and most of the stroller materials can be recycled too.

Particularly compelling is the categorization of parts acording to the number of potential (re)use cycles in a circular economy. The first category are the lasting ones, engineered to function the full lifespan, but with the potential to go through more use cycles. They consist of durable metals, composites and high-grade plastics. Category two contains parts that can wear and tear and last five years minimum. Wheels and small plastic clamps belong here. Next there are parts for refurbishment and fresh looks, such as the textile ones. The final category is packaging, which is entirely made out of cardboard, with the exception of one plastic bag for cushions, because they have to be kept dry.

The company is in touch with its second-hand market and monitors use by parents, and their friends and their family and so on. A rather unique initiative are partnerships with companies like Atelier Little Boomerang, that repurposes leftover materials from strollers into carrier bags. Looking for this kind of cooperation is crucial for circular management and prepares the company for an even more responsible future.

EGG HELMETS The egg is a highly customizable multi-sport helmet for kids that was designed from the onset with the goal of matching a high level of fun with a high level of safety and comfort. Colourful skins can easily be combined with a choice of add-ons allowing users to match the look of the helmet with their moods. A safety 'must' thus becomes a fashion 'want'.

OPEN SOURCE CINEMA CAMERA *Build your dream camera, one module after the other. AXIOM is an open hardware plus free software family digital of cinema camera devices. AXIOM camera systems are not only fully open but also built to be entirely modular, ready to evolve into things not even imaginable today. The Open Module Concept renders changing camera parts as easy as exchanging PCIe cards in a PC. It therefore renders AXIOM cameras to be fully extendible and keeps them from being outdated any time soon. This ensures sustainability of the entire assembly, since any module - such as the image sensor - can be replaced while keeping the rest of the camera intact.*

MANHATTAN RESEARCH Raymond Scott was an American composer, band leader, pianist, engineer, recording studio maverick, and electronic instrument inventor. In 1946, Scott established Manhattan Research, which he announced would 'design and manufacture electronic music devices and systems.' With his company Scott designed audio devices for his own personal use and provided customers with sales & service for a variety of devices, for the creation of electronic music and 'musique concrete' including components such as ring modulators, wave, tone and envelope shapers, modulators and filters. And instruments such as 'Chromatic electronic drum generators' and 'Circle generators'. He called himself the inventor of the polyphonic sequencer. Scott often described Manhattan Research Inc. as 'More than a think factory - a dream center where the excitement of tomorrow is made available today.'

Scott developed some of the first devices capable of producing a range of electronic tones automatically in sequence. He began working on a machine that accorrding to him made compositions with artificial intelligence. Scott called it the Electronium. With its vast array of switches, buttons and patch panels it is considered to be the first self-composing synthesizer.

Some of Raymond Scott's projects were less complex, but still ambitious. During the 1950s and 1960s he produced electronic telephone ringers, alarms, chimes, and sirens, vending machines and ashtrays with accompanying electronic music scores, and an adult toy that produced sounds that varied depending on how two people touched one another.

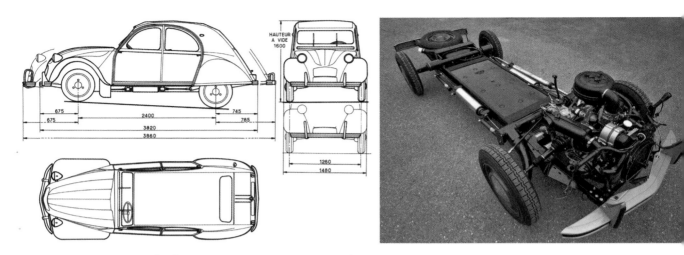

MAINTANANCE AND (DIS)ASSEMBLY *It has been described as 'the most intelligent application of minimalism ever to succeed as a car. The name in French: 'Deux Chevaux Vapeur', in short 2CV. The name translated into English: 'two steam horses'. It is a remarkably simple and practical car. The first version even had just a single headlamp and one windscreen wiper. It was meant to be a cheap motorized vehicle for farmers to conveniently transport pigs, or straw, or family. It featured low cost, simple maintenance, an accessible air cooled engine and low fuel consumption. Exceptional torsion suspension offered a smooth driving experience. It was destined to become popular among students and intellectuals. Citroën manufactured 8.7 million of them.*

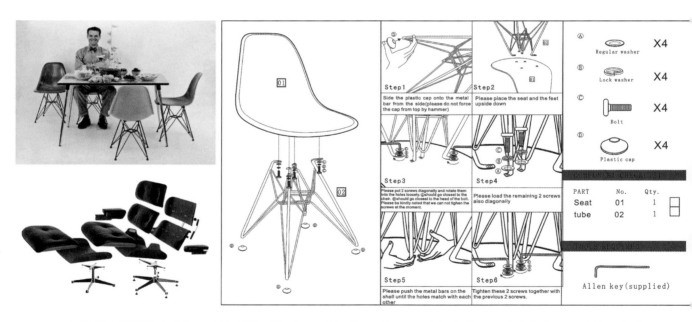

FOR THE MASSES *Post-Second World War, Charles and Ray Eames (he an architect, she an artist, designer and filmmaker) created ground breaking contributions to a variety of disciplines: architecture, furniture design, industrial design, manufacturing and photography. Their philosophy was a democratic one, not wanting to appeal to the rich elitists but to the masses, wanting their designs to deliver 'the best to the greatest number of people for the least'. That low cost approach to furniture with smart designs that could be shipped and stored in compact boxes and could be assembled by the customers themselves, has been adopted by many – most notably that Swedish flat pack giant. Eames furniture itself has somehow turned elitist.*

About half a century ago, there was a student who lived on the third floor of a student apartment building. He was the very proud owner of a Citroen 2CV, a small and straightforward cheap car, very popular among students in those days and nicknamed 'Duck'. He went on holiday and asked a friend to take good care of it, which the friend did, in a very special way. Together with a group of fellow students he took the car apart and reassembled it in the student's room on the third floor. Now try and imagine this guy's face when he returned home from abroad and found his car in his room.

DIS- AND REASSEMBLY

Some products are designed with this kind of handling in mind. Probably the most popular bed in Europe is Auronde by Auping. Frans de la Haye designed it in 1973 and it is still going strong. No doubt it is a classic, with a quality that nourishes attachment. There have been upgrades in technical details for adjustment and in respect of the availability of colours, but in principle the system has remained the same. Apart from its restrained looks and adequate functionality, its success is also caused by the fact that it can be taken apart, put together again, and has exchange components. A single bed can become a double bed and a low bed can be made higher.

Age does not matter, parts will always fit. Here we have a superb demonstration of the lifespan advantage of being able to disassemble a product and put it together again, and again, and again.

In this respect the bed is not entirely unique. There are office desk systems similar to it, such as Mehes by Ahrend, designed by Friso Kramer in 1974. Landrovers have this kind of quality as well, as does the Bailey bridge. This is a simple and light segmented system to build a bridge up to 60 metres long for heavy army traffic. After use it can be dismantled and stored for reuse. A segment is light enough to be handled by a few men. No cranes or special tools are required. It was developed in 1943 and is still in use.

By far the most successful assault weapon is Kalashnikov's AK47. Approximately 75 million of these were produced. Its simplicity has made it into a huge success. Anything can be done to it and it will still work. It will shoot after it has been run over by a truck

and it will shoot under water (not very far). The record set for disassembly and reassembly, the so-called 'field strip' stands at around 13 seconds. The army apparently favours products designed for easy disassembly and reassembly.

The connection of disassembly with reassembly provides an important difference from the kind of disassembly defined for eco-design, where it does not really matter how parts are put together, as long as you can take them apart.

The best way to understand the requirements for disassembly and reassembly is to make a comparison with products that are not very good in this respect. This is a matter of detail. When you buy a wardrobe at IKEA, it comes in boxes with parts that have to be put together. For the sake of this argument, it does not matter whether this is difficult or not. The main point is that the back of the wardrobe consists of a sheet of fibreboard that must be nailed (with small nails) to the back of the top, bottom and side panels that have been bolted together. These consist of chipboard. Taking the whole thing apart is easy with hammers and fists, but suppose you need to put it back together again: the nails will have damaged the fibreboard and the chipboard as nails do, particularly to the chipboard. The components lose their constructive integrity.

For disassembly and reassembly, it is crucial that every part is an independent product and has to be designed and tested as such. It carries its value in an elegant way. This does not rule out gluing, nailing and riveting altogether, as long as it does not interfere with the disassembly and reassembly procedure. Glue a rubber strip to a metal panel and it will become part of it forever.

ARCHITECTURE AND (DIS)ASSEMBLY *Architecture is sometimes a silly profession. It takes a very long time and a lot of money to design every house as a one-off prototype, and that makes no sense if you have a really good replicable design. Flat boxed houses are produced in series, shipped in boxes and can be assembled, disassembled and replaced. Be it permanent housing or temporary shelters.*

FLAT PACK HOUSES Kiss House comes in two, three and four bedroom versions. Configurable elements enable clients to make kiss house their own. The house is built according to passive energy principles, out of cross-laminated timber (CLT) panels. All Kiss houses are triple glazed. They can be customized using various external finishes and colour scemes. The house is flat packed so that it can be shipped efficiently and assembled very quickly.

REFUGEE SHELTER Better Shelter is a collaboration between the IKEA Foundation and United Nations High Commissioner for Refugees (UNCHR). The Better Shelter comes in flat packed cardboard boxes and people can start building it immediately by following an instruction manual. It is constructed in three subsequent steps: the foundations first, then the roof that includes ventilation and the solar panel, finally the walls with windows and a door. It takes four people four to eight hours to complete the construction. No additional tools are required and most components are assembled by hand.

At a cost of $1,150 each, the shelters are three times more expensive than a standard UNHCR tent. Yet while the latter are designed to last for just six months, these new shelters last for a minimum of three years in harsh conditions, and up to twenty years in more temperate climates.

BUILDINGS BECOME MATERIAL BANKS In the next 40 years, we will be building just as much as we have built so far in human history. This requires a radically different mindset. Tomorrow's buildings must be designed for disassembly to ensure component and material contin-use. London-based design studio LYN Atelier designed a temporary community center in Hackney Wick, constructed using recycled materials from the London 2012 Olympic and Paralympic Games. The community center, named Hub 67, has a lifespan of up to five years, after which it will be disassembled and its parts reused elsewhere.

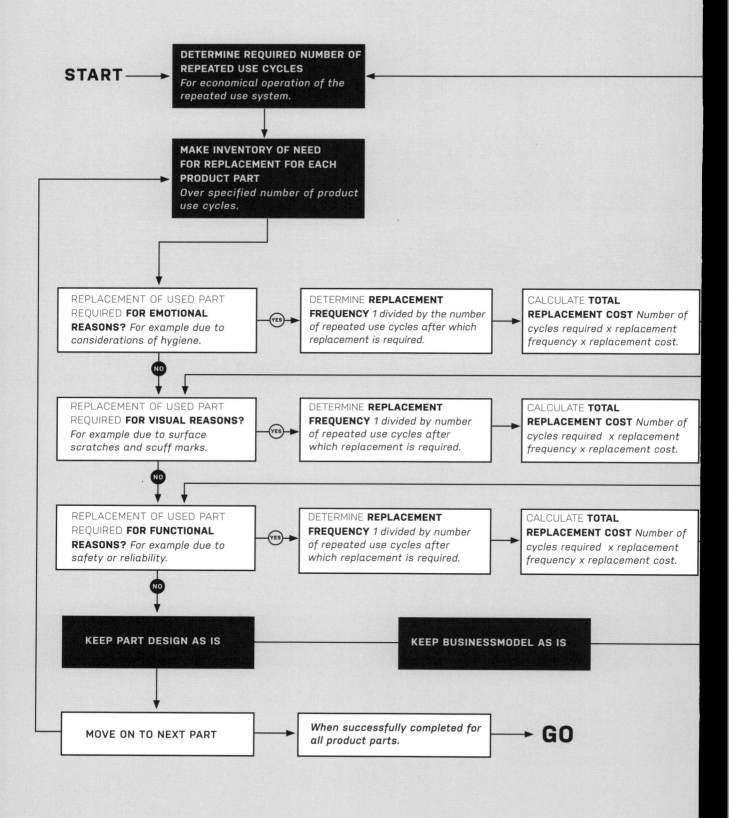

START ➞ **DETERMINE REQUIRED NUMBER OF REPEATED USE CYCLES** *For economical operation of the repeated use system.*

MAKE INVENTORY OF NEED FOR REPLACEMENT FOR EACH PRODUCT PART *Over specified number of product use cycles.*

REPLACEMENT OF USED PART REQUIRED **FOR EMOTIONAL REASONS?** *For example due to considerations of hygiene.*

YES ➞ DETERMINE **REPLACEMENT FREQUENCY** *1 divided by the number of repeated use cycles after which replacement is required.*

CALCULATE **TOTAL REPLACEMENT COST** *Number of cycles required x replacement frequency x replacement cost.*

NO

REPLACEMENT OF USED PART REQUIRED **FOR VISUAL REASONS?** *For example due to surface scratches and scuff marks.*

YES ➞ DETERMINE **REPLACEMENT FREQUENCY** *1 divided by number of repeated use cycles after which replacement is required.*

CALCULATE **TOTAL REPLACEMENT COST** *Number of cycles required x replacement frequency x replacement cost.*

NO

REPLACEMENT OF USED PART REQUIRED **FOR FUNCTIONAL REASONS?** *For example due to safety or reliability.*

YES ➞ DETERMINE **REPLACEMENT FREQUENCY** *1 divided by number of repeated use cycles after which replacement is required.*

CALCULATE **TOTAL REPLACEMENT COST** *Number of cycles required x replacement frequency x replacement cost.*

NO

KEEP PART DESIGN AS IS

KEEP BUSINESSMODEL AS IS

MOVE ON TO NEXT PART ➞ *When successfully completed for all product parts.* ➞ **GO**

PRODUCTS THAT LAST FINAL CHECKLIST This flow diagram will help you determine whether your product or a product design is suitable for application in a business model that relies on the product being used by a number of different users. Each user will expect a safe, fully functional and reasonably pristine product. This tool provides a schematic overview of the steps needed to perform such a check.

Intangible, or emotional, arguments for part replacement prevail over aesthetical, or visual, arguments. Functional, reliability and –most important- safety, arguments form the final and all-important go/no go hurdle each part of the product has to take. Too frequent part replacements can lead to high operational costs, making a business model not feasible from a financial point of view. Matching business model, part and product design is essential in order to achieve a robust and profitable business solution.

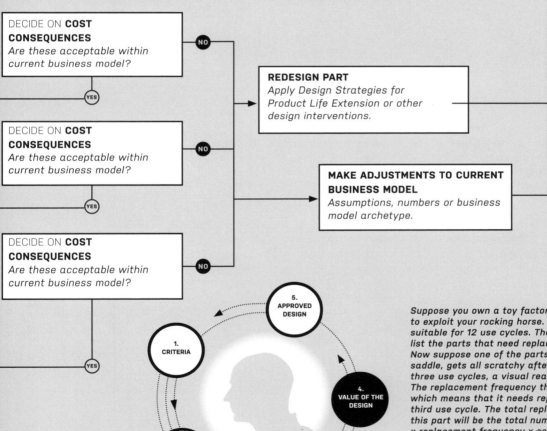

DECIDE ON **COST CONSEQUENCES**
Are these acceptable within current business model?

NO **YES**

DECIDE ON **COST CONSEQUENCES**
Are these acceptable within current business model?

NO **YES**

DECIDE ON **COST CONSEQUENCES**
Are these acceptable within current business model?

NO **YES**

REDESIGN PART
Apply Design Strategies for Product Life Extension or other design interventions.

MAKE ADJUSTMENTS TO CURRENT BUSINESS MODEL
Assumptions, numbers or business model archetype.

5.
APPROVED DESIGN

1.
CRITERIA

4.
VALUE OF THE DESIGN

2.
PROVISIONAL DESIGN

3.
EXPECTED PROPERTIES

Suppose you own a toy factory and you want to exploit your rocking horse. You want it to be suitable for 12 use cycles. Then first you have to list the parts that need replacing and how often. Now suppose one of the parts, a 10 Euro plastic saddle, gets all scratchy after, for instance, three use cycles, a visual reason to replace it. The replacement frequency then would be 1/3, which means that it needs replacing after every third use cycle. The total replacement costs for this part will be the total number of use cycles x replacement frequency x cost = 12 x 1/3 x 10 = 40 Euros. Within the overview of the complete exploitation this will provide a way to decide on feasibility. The design and the business model may need adjustment.

OUR TASK *Machines can perform boring and dangerous tasks for years on end. They don't really care, about us, or anything at all. Managing the effect of our skills is not up to technology, but up to us.*

RECOMMENDED READING

Please note: Some of the books in this list are quite old and have become hard to find and rather expensive. Should you be able to get your hands on them, don't hesitate, for chances are they too will gain in value.

ON BUSINESS & STRATEGY

BUSINESS MODEL GENERATION. Alexander Osterwalder and Yves Pigneur, 2010. John Wiley and Sons, ISBN 978-047-087641-1

STRATEGY: *Formulating Analytical Concepts.* Charles W. Hofer and Dan Schendel, 1978. West Publishing Company, ISBN 0-8299-0213.

STRATEGY: *Seeking and Securing Competitive Advantage.* Cynthia A. Montgomery and Micheal E. Porter (eds.) 1991. Harvard Business Press, ISBN 0-87584-243-7

ON VALUE AND CREATING VALUE PROPOSITIONS

MAKING MEANING: *How Succesful Businesses Deliver Meaningful Customer Experiences.* Steve Diller, Nathan Shedroff and Darrel Rhea, 2008. New Riders, ISBN 0-321-55234-2

RUBBISH THEORY Micheal Thompson, 1979. Oxford University Press, ISBN 0-19-217658-7

VALUE PROPOSITION DESIGN: *How to Create Products and Services Customers Want.* Alexander Osterwalder, Yves Pigneur, Gregory Bernarda, Alan Smith and Trish Papadakos, 2014. John Wiley and Sons, ISBN 978-1-118-96805-5

ZEN AND THE ART OF MOTORCYCLE MAINTENANCE. Robert M. Pirsig, 1975. Bantham Books, ISBN 978-0-553-08880-9

ON PLANNED OBSOLESCENCE & CONSUMERISM

CHEAP: *The High Cost of Discount Culture.* Ellen Ruppel Shell, 2010. Penguin Books, ISBN 978-0143117636.

ENDING THE DEPRESSION THROUGH PLANNED OBSOLESCENCE. Bernard London, 1932. https://upload.wikimedia.org/wikipedia/commons/2/27/London_(1932)_Ending_the_depression_through_planned_obsolescence.pdf

ENOUGH. John Naish, 2008. Hodder & Stoughton, ISBN 978-0-340-93592-7

HIDDEN PERSUADERS. Vance Packard, 1957. David McKay.

MADE TO BREAK. *Technology and Obsolescence in America.* Giles Slade, 2006. Harvard University Press, ISBN 978-0-674-02572

THE OVERSPENT AMERICAN. *Why We Want What We Don't Need.* Juliet B. Schor Harper Collins, ISBN 978-0-06-097758

THE PARADOX OF CHOICE: *Why More is Less.* Barry Schwarz, 2005. Harper Perennial, ISBN 978-0060005696.

THE STORY OF STUFF: *The Impact of Overconsumption on the Planet, Our Communities, and Our Health-And How We Can Make It Better.* Annie Leonard, 2011. Free Press, ISBN 978-1-451-61029-1

WASTE MAKERS. Vance Packard, 1960. David McKay.

ON PRODUCT DESIGN & DESIGN METHODS

CHANGE BY DESIGN; *How design thinking transforms organizations and inspires innovation.* Tim Brown, 2009. HarperCollins Publishers. ISBN 978-0-06-176608-4

DESIGN FOR MANAGING OBSOLESCENCE; *A Design Methodology for Preserving Product Integrity in a Circular Economy.* Marcel den Hollander, 2018. Marcel den Hollander IDCR, ISBN 978-90-8287-360-3

101 DESIGN METHODS: *A Structural Approach for Driving Innovation in Your Organization.* Vijay Kumar, 2013. John Wiley & Sons, ISBN 978-1-118-08346-8

DELFT DESIGN GUIDE. Annemiek van Boeien, Jaap Daalhuizen, Jelle Zijlstra and Roos van der Schoor, 2013. Bis Publishers, ISBN 978-90-6369-327-5

DESIGN FOR THE REAL WORLD: *Human Ecology and Social Change* – second revised edition. Victor Papanek, 2005. Chicago Review Press, ISBN 978-0-897-33153-1

GLIMMER: *How Design Can Transform Your World.* Warren Berger and Bruce Mau, 2010. Vintage Canada, ISBN-13: 978-03-0735 674-1

HOW TO GET IDEAS. Jack Foster, 2007. Berrett-Koehler Publishers Inc., ISBN 978-1-57675-430-6

THE LANGUAGE OF THINGS: *Understanding the World of Desirable Objects.* Deyan Sudjic, 2009. W. W. Norton & Company, ISBN 978-0-393-070811

THE SOLID SIDE; *the search for consistency in a changing world; projects and proposals.* Ezio Manzini and Marco Susani (eds.) Domus Academy & Philips Corporate Design, 1995. V&K Publishing, the Netherlands.

ON (INFLUENCING) HUMAN BEHAVIOUR & MOTIVATION

DRIVE. *The Surprising Truth About What Motivates Us.* Daniel H. Pink, 2009. Cannongate, ISBN 978-1-84767-769-3

NUDGE. IMPROVING DECISIONS ABOUT HEALTH, WEALTH, AND HAPPINESS. Richard H. Thaler and Cass R. Sunstein, 2008. Yale University Press/Caravan, ISBN 978-0-300-12223-7

STUCK. *Why We Can't (or Won't) Move On.* Ameli Rufus, 2009. Capstone Publishing Ltd, ISBN 978-1-907-31245-8

CLOCK OF THE LONG NOW: *Time And Responsibility: The Ideas Behind The World's Slowest Computer.* Steward Brand, 2000. Basic Books, ISBN 978-0-465-00780-6

ON SYSTEMS THINKING & MANAGEMENT

THE FIFTH DISCIPLINE: *The Art & Practice of The Learning Organization.* Peter Senge, 2006. Doubleday, ISBN 978-0-385-51725-6.

THE GOAL. Eliyahu M. Goldratt and Jeff Cox, 2004. North River Press, ISBN 978-0-884-27178-9

ON ALTERNATIVE OPTIONS FOR BUSINESS, CONSUMPTION AND ECONOMY

DOUGHNUT ECONOMICS. Kate Raworth, 2018. Random House. ISBN 9781847941398

LONGER LASTING PRODUCTS; *Alternatives to the Throwaway Society.* Tim Cooper (ed), 2010. Gower, ISBN 978-0566088087

THE AGE OF ACCESS. Jeremy Rifkin, 2000. Tarcher/Putnam, ISBN 1-58542-082-4.

THE DURABLE USE OF CONSUMER PRODUCTS Michel Kostecki (ed.), 1998. Kluwer Academic Publishers, ISBN 0-7923-8145-9.

THE PERFORMANCE ECONOMY-SECOND EDITION. Walter R. Stahel, 2010. Palgrave Macmillan, ISBN 978-0-230-58466-2.

BEYOND ECONOMICS AND ECOLOGY: *the radical thought of Ivan Illich.* Ivan Illich (edited by S. Samuel and J. Brown), 2013. Marion Boyars Publishers Ltd, ISBN 978-0-714-53158-8

WHAT'S MINE IS YOURS. *How Collaborative Consumption Is Changing The Way We Live* Rachel Botsman and Roo Rogers, 2010. Harper Collins, ISBN 978-0-00-739591-0

PROSPERITY WITHOUT GROWTH: ECONOMICS FOR A FINITE PLANET. Tim Jackson, 2011 Routledge, ISBN 978-1-849-71323-8

ON OUR PREDICAMENT

THE MUSHROOM AT THE END OF THE WORLD; *on the possibility of life in capitalist ruins.* Anna Lowenhaupt Tsing, 2015, Princeton University Press. ISBN 978-0-691-17832-5

AFTER PROGRESS; *Reason and Religion at the End of The Industrial Age.* John Michael Greer, 2015. New Society Publishers. ISBN 978-0-86571-791-6

AN INCONVENIENT TRUTH. Al Gore, 2006. Rodale Books, 978-1-594-86567-1

DEAD MEDIA MANIFESTO Bruce Sterling www.deadmedia.org/modest-proposal.html

EAARTH: *Making Life on a Tough New Planet.* Bill McKibben, 2011. St. Martin's Griffin, ISBN 978-0312541194

OUR ANGRY EARTH. Isaac Asimov and Frederik Pohl, 1991. Tor Books, ISBN 978-0-312-85252-8

THE INGENUITY GAP: *Can We Solve The Problems of the Future?* Thomas Homer-Dixon, 2000. Vintage Canada, ISBN 0-676-97296-9

THE NEXT 100 YEARS: *A Forecast for the 21st Century.* George Friedman, 2010. Anchor, ISBN 978-0767923057

THE STORY OF B: *An Adventure of the Mind and Spirit.* Daniel Quinn, 1996. Bantam, ISBN 978-0-553-37901-3

ON CLOSED LOOP SYSTEMS & INDUSTRIAL ECOLOGY

BIOMIMICRY, *Innovation Inspired by Nature.* Janine Benyus, 2002. Perennial. ISBN 0-688-16099-9

CRADLE TO CRADLE: *Remaking the Way We Make Things.* Micheal Braungart and William McDonough, 2002. North Point Press, ISBN 978-0-865-47587-8

INDUSTRIAL ECOLOGY – SECOND EDITION T.E. Graedel and B.R. Allenby, 2003. Pearson Educaction (Prentice Hall), ISBN 9-780-1-3046713-3

REGENERATIVE DESIGN FOR SUSTAINABLE DEVELOPMENT. John Tillman Lyle, 1994. John Wiley & Sons, ISBN 9-780-471-55582-7

TOWARDS THE CIRCULAR ECONOMY, reports 1, 2 & 3. Ellen McArthur Foundation, 2012, 2013 and 2014. www.ellenmacarthurfoundation.org/business/reports

COLOPHON

© 2019 Conny Bakker, Marcel den Hollander, Ed van Hinte, Yvo Zijlstra and BIS Publishers

BIS Publishers - Borneostraat 80-A - 1094 CP Amsterdam - The Netherlands
T +31 (0)20 515 02 30 bis@bispublishers.com / www.bispublishers.com

2nd printing 2022

ISBN 978-90-6369-522-4

Netherlands Enterprise Agency

The PRODUCTS THAT LAST project was supported by the Netherlands Enterprise Agency (RVO) as part of the Innovation-Oriented Research Programme 'Integrated Product Creation and Realisation (IOP IPCR)' of the Ministry of Economic Affairs.

ABOUT THE AUTHORS

This book emerged from an inspiring and synergetic creative process between senior researcher and industrial designer Marcel den Hollander (drmarceldenhollander@gmail.com), professor Conny Bakker (c.a.bakker@tudelft.nl), writer Ed van Hinte (ejhint@wxs.nl) and graphic designer Yvo Zijlstra (zijlstra@antenna-men.com). Copy editor: Roger Staats.

IMAGE CREDITS p6. Boy with football in Jakarta, Indonesia, Jonathan McIntosh, Creative Commons p8. ESA's Clean Space Office - p18. Michael Kooren, Hollandse Hoogte - p16-17. Picturenarrative / Buzludzhap p 22. Saltspring Coffee Roasting Facility by Kris Krug, creative commons - p 23. Chris McClanahan - p 25. Utility Fashion designed by Norman Hartnell p26 Wikimedia CC; Jacob Owens; Aerodrums - p27. budnitzbicycles. com p29. Shadow Robot Co. - p30. Image courtsesy Hiroshi Fuji - p32 courtesy Riversimple; Wikimedia CC - p34. Steve Parsons - www.sebrightcreative.co.uk - p36. The Long Now Foundation, Rolfe Horn - p38. Commons Wikimedia CC- p40. Shamees Aden - P42. Commons Wikimedia - p44. Wikimedia ; Abfad - p46-47. Bioglow, Dan Saunders - Agricultural Research Service of the United States Department of Agriculture; Softskill Design; Lego - p73. Source: rowantechnology - p53. Wikimedia CC - p 56. Carlos Jones; Oak Ridge National Lab - p59. Wikimedia; IBM; TU Delft/QuTech - p70. Vitsoe - p88. Wikimedia CC; Swapfiets - p 91 Courtesy Umicore - p 92 Wikimedia CC- p96. Wikimedia CC - p98. Patek Philipe; Marcel den Hollander - p100. T.J. Huff; Teimen Dzjind - p102. Bauknecht - p104-105. Airbus SAS; S. Ramadier, H. Goussé, all rights reserved - p106. Golan Levin; Joshua Silver - p109. images courtesy BBC & Plum Pictures - p112. Marvin A. Moss, 'Designing for Minimal Maintenance Expense', 1985. Marcel Dekker Inc., New York - p115. University of Illinois; University of Stanford; Nissan Motor - p114. Fairphone - p116. Bugaboo; Egg Helmet - p117. Axiom; Wikimedia - p120. Kisshouse.co.uk; Jonas Nyström/ Better Shelter; LYN atelier - p124. Shutterstock